ENVIRONMENTAL SCIENCE, ENGINEERING AND TECHNOLOGY

ENVIRONMENTAL CONSIDERATIONS IN AGRICULTURE

ENVIRONMENTAL SCIENCE, ENGINEERING AND TECHNOLOGY

Additional books in this series can be found on Nova's website under the Series tab.

Additional E-books in this series can be found on Nova's website under the E-books tab.

AGRICULTURE ISSUES AND POLICIES

Additional books in this series can be found on Nova's website under the Series tab.

Additional E-books in this series can be found on Nova's website under the E-books tab.

ENVIRONMENTAL SCIENCE, ENGINEERING AND TECHNOLOGY

ENVIRONMENTAL CONSIDERATIONS IN AGRICULTURE

ERIC A. SARGSYAN
AND
ADAM J. OCHION
EDITORS

Nova Science Publishers, Inc.
New York

For permission to use material from this book please contact us:
Telephone 631-231-7269; Fax 631-231-8175
Web Site: http://www.novapublishers.com

Additional color graphics may be available in the e-book version of this book.

Library of Congress Cataloging-in-Publication Data

Environmental considerations in agriculture / editors, Eric A. Sargsyan and Adam J. Ochion.
p. cm.
Includes bibliographical references and index.
ISBN 978-1-61470-077-7 (hardcover)
1. Agriculture--Environmental aspects--United States. 2. Environmental policy--United States. 3. Agriculture and state--United States. I. Sargsyan, Eric A. II. Ochion, Adam J.
S589.755.E57 2011
630'.286--dc23

2011020184

Published by Nova Science Publishers, Inc. † New York

CONTENTS

PREFACE

Some members of Congress have expressed concern about recent environmental regulations and administrative initiatives. Criticism from lawmakers and industry leaders is primarily focused on environmental regulations promulgated by the Environmental Protection Agency (EPA). Some claim that EPA is overreaching its regulatory authority in several environmental arenas. The agriculture community has been vocal with its concerns, contending that EPA appears to be focusing its regulatory efforts on agriculture. Environmentalists, on the other hand, are encouraged by some of EPAs regulatory efforts, claiming that some agriculture operations do pose a public health and environmental risk and should be regulated. This book examines select environmental regulations that could affect agriculture.

Chapter 1- Some in Congress have expressed concern about recent environmental regulations and administrative initiatives. Criticism from lawmakers and industry leaders is primarily focused on environmental regulations promulgated by the Environmental Protection Agency (EPA). Some claim that EPA is overreaching its regulatory authority in several environmental arenas. The agriculture community has been vocal with its concerns, contending that EPA appears to be focusing its regulatory efforts on agriculture. Environmentalists, on the other hand, are encouraged by some of EPAs regulatory efforts, claiming that some agriculture operations do pose a public health and environmental risk and should be regulated.

Chapter 2- This report examines the implications for agriculture of the ongoing but inconclusive debate about global climate change. In that debate, agriculture's role is multifaceted. Agriculture is both a source of several greenhouse gases (GHGs) and a "sink" for absorbing carbon dioxide, the most common GHG, thereby partly offsetting emissions. Overall, agriculture is a comparatively modest source of U.S. GHG emissions: it accounts for approximately 7% of U.S. emissions, while sectors such as transportation and electricity generation account for much larger shares. Agriculture's GHG emissions are principally in the form of methane and nitrous oxides emissions.

Chapter 3- The agriculture sector is a source of greenhouse gas (GHG) emissions, which many scientists agree are contributing to observed climate change. Agriculture is also a "sink" for sequestering carbon, which might offset GHG emissions by capturing and storing carbon in agricultural soils. The two key types of GHG emissions associated with agricultural activities are methane (CH_4) and nitrous oxide (N_2O). Agricultural sources of CH_4 emissions mostly occur as part of the natural digestive process of animals and manure management at livestock operations; sources of N_2O emissions are associated with soil management and fertilizer use on croplands. This report describes these emissions on a carbon-equivalent basis to illustrate agriculture's contribution to total national GHG emissions and to contrast emissions against estimates of sequestered carbon.

Chapter 4- Gases other than carbon dioxide accounted for nearly 15% of total U.S. greenhouse gas emissions in 2008, yet there has been minimal discussion of these other greenhouse gases in climate and energy legislative initiatives. Reducing emissions from non-carbon dioxide greenhouse gases, such as nitrous oxide (N_2O), could deliver short-term climate change mitigation results as part of a comprehensive policy approach to combat climate change.

Chapter 5- Agriculture could play a prominent role in U.S. efforts to address climate change if farms and ranches undertake activities that reduce greenhouse gas (GHG) emissions or take greenhouse gases out of the atmosphere. These activities may include shifting to conservation tillage, reducing the amount of nitrogen fertilizer applied to crops, changing livestock and manure management practices, and planting trees or grass. The Federal Government is considering offering carbon offsets and incentive payments to encourage rural landowners to pursue these climate-friendly activities as part of a broader effort to combat climate change. The extent to which farmers adopt such activities would depend on their costs, potential revenues, and other economic incentives created by climate policy. Existing Federal conservation programs provide preliminary estimates of the costs of agri-cultural carbon sequestration.

Chapter 6- If agricultural producers are allowed to participate in a national cap-and-trade system to curb greenhouse gas (GHG) emissions, the opportunity to sell carbon off sets could prompt farmers to manage their land in a way that increases the amount of carbon stored in soil organic matter and plant biomass, including residue. (See box, "Agriculture and Carbon Off sets.") Farmers who own the land they farm, however, may be in a better position to generate off sets than those who rent their land.

In: Environmental Considerations in Agriculture ISBN: 978-1-61470-077-7
Editors: E.A. Sargsyan, A.J. Ochion 2011 Nova Science Publishers, Inc.

Chapter 1

ENVIRONMENTAL REGULATION AND AGRICULTURE*

Megan Stubbs

SUMMARY

Some in Congress have expressed concern about recent environmental regulations and administrative initiatives. Criticism from lawmakers and industry leaders is primarily focused on environmental regulations promulgated by the Environmental Protection Agency (EPA). Some claim that EPA is overreaching its regulatory authority in several environmental arenas. The agriculture community has been vocal with its concerns, contending that EPA appears to be focusing its regulatory efforts on agriculture. Environmentalists, on the other hand, are encouraged by some of EPAs regulatory efforts, claiming that some agriculture operations do pose a public health and environmental risk and should be regulated.

A healthy agriculture industry and a healthy environment are both important to the nation. However, agricultural production can have varying impacts on the environment. The use of both natural resources and synthetic inputs in agricultural production can sometimes create a negative impact on human health and the surrounding ecosystem. The magnitude of these environmental impacts vary widely across the country and change over time. Given the agricultural sector's size and potential to

---

* This is an edited, reformatted and augmented version of a Congressional Research Service 7-5700, R41622 publication, dated February 7, 2011.

affect its surrounding environment, there is interest in both tightening environmental policies while also maintaining an economically viable industry. Most recognize the agriculture community's efforts to protect natural resources while striving to maintain a sustainable and abundant food supply.

The current federal response to environmental issues associated with agriculture is viewed as being both restrictive and supportive. Traditionally, most farm and ranch operations have been exempt or excluded from many environmental regulations. The challenges associated with regulating numerous crop and livestock operations, can be cost prohibitive for government regulators, and environmental policies have historically focused on large industrial sources such as factories and power plants. Therefore, much of the current farm policy addressing environmental concerns is in the form of economic incentives to encourage beneficial production practices.

Recent regulatory activity has generated widespread interest in the depth of EPA's regulatory authority. The 112th Congress may evaluate EPA and other federal agencies' roles in regulating environmental protection generally. Other broad options for Congress besides general oversight include review under the Congressional Review Act, amending current law to modify a regulating agency's authority, introducing freestanding legislation, or offering an amendment on an agency's appropriation bill that prevents funds from being used for specific actions.

This report covers select environmental regulations that could affect agriculture. The majority of environmental regulations are administered by EPA, though not all. In some cases, agriculture is the direct or primary focus of the regulatory actions. In other cases, the agriculture sector is one of many affected sectors. Of particular interest to the sector are regulatory actions affecting air, water, energy and chemicals. Issues associated with air (e.g., dust and emission) and water (e.g., fertilizer and nutrient run-off) resources are a primary focus to many regulations affecting agriculture because of agriculture's potential impact to both. Changes in energy policy, namely bioenergy, have recently become important to many in the agricultural industry based on the growing influence of corn-based biofuel production. Finally, the risks associated with agricultural chemical use and possible impacts on human health and the environment have led to recent federal regulatory reviews of chemical fertilizer and pesticide use.

INTRODUCTION

A healthy agriculture industry and a healthy environment are both important to the nation. However, agricultural production can have varying impacts on the environment. The use of both natural resources (e.g., soil and water) and synthetic inputs (e.g., fertilizers and pesticides) in agricultural production can sometimes create a negative impact on the surrounding ecosystem. For example, soil erosion, farm chemical runoff, and overgrazing can affect water and air resources. Converting grassland prairies and wetlands to crop production can impact wildlife populations. The magnitude of these environmental impacts vary widely across the country and change over time.

The federal response to agriculture's impact on the environment can be viewed at opposite ends of a spectrum: incentivizing sustainable production (carrot) versus requiring it through regulation (stick). While most within the agriculture community prefer the "carrot" approach, there is an increasing focus on the "stick" because of recent federal regulatory action. Current federal environmental policies both restrict and encourage certain production practices. The ultimate mix of policy instruments depends on the nature of the resource issue and the information available on the linkages between farming activities and the environmental resources.

Traditionally, most farm and ranch operations have been exempt or excluded from many federal environmental regulations. Attempting to regulate numerous individual crop and livestock operations can be cost prohibitive for government regulators, and environmental policies have historically focused on large industrial sources such as factories and power plants. Therefore, much of the current federal farm policy addressing environmental concerns is in large part voluntary, that is, seeking to encourage agriculture producers to adopt conservation practices through economic incentives. Most environmental regulation, in terms of permitting, inspection and enforcement, is done by state and local governments, typically based on federal EPA regulatory guidance. Many point out that the relative number of environmental regulations affecting agriculture are few compared to other industries. However, given the agricultural sector's size in the landscape[1] and its potential to affect its surrounding environment, there is interest in both tightening environmental policies while also maintaining an economically viable agriculture industry.

The U.S. Environmental Protection Agency (EPA) is the primary federal authority for administering environmental protection polices, while the U.S. Department of Agriculture (USDA) is the primary federal authority for incentivizing agricultural production. USDA provides both educational outreach, and technical and financial assistance opportunities for producers to implement environmentally sustainable practices.[2] And while many of these voluntary programs and policies have been in place for decades and have had considerable success, some question whether a strictly voluntary approach to agricultural conservation generates enough environmental gains.[3] EPA, on the other hand, has recently received criticism from lawmakers and industry leaders for appearing to focus regulatory efforts on agriculture.[4] Some claim EPA has overreached its regulatory authority.[5] Some in the agriculture community have been vocal in their displeasure with recent EPA regulatory proposals and the costs associated with providing a more sustainable production system.[6]

The 112th Congress will likely give attention to EPA's and other federal agencies' roles in regulating environmental protection generally. Both the Senate and House Committees on Agriculture have shown particular interest in EPA's actions and conducted oversight hearings on regulatory impacts to agriculture during the 111th Congress; similar activity is expected during the 112th Congress.[7]

Report Content and Caveats

This report covers select federal environmental regulations that could affect agriculture.[8] This report is intended to provide the background, status, and issues related to environmental regulations or initiatives possibly affecting agriculture. Many of these issues are commonly referred to as being of concern to agriculture based on media and industry reports. Their inclusion in this report is not intended to suggest or imply that the regulation or action has either a beneficial or harmful effect on agriculture or to what degree. Similarly, regulatory actions not included in this report do not indicate the lack of potential impact on the agriculture sector.

This report only addresses federal regulatory actions. In many cases, constraints on agricultural production to reduce pollution emissions arise at the state level in response to local concerns. State and local regulations are not specifically included in this report, but may be discussed generally where appropriate. Actions considered voluntary or in response to regulatory actions are also not included. This means that many USDA programs and initiatives, which offer funding to agricultural producers to preclude the need for environmental regulation, are not discussed in this report.

The majority of the regulations discussed in this report are administered by EPA, though not all. In some cases, agriculture is the direct or primary focus of the regulatory actions. In other cases, agriculture is one of many affected sectors. In many cases for a regulation to become effective, EPA rules must be adopted by states to which the program has been delegated (e.g., most environmental permitting programs are delegated to qualified states). Moreover, many states require that the state legislature review new regulations before the new rules would take effect. The general regulatory development and compliance process can be tedious and complex. In some cases the promulgation and implementation of regulations may take years.[9] In the case of some environmental regulations, the agencies must adhere to court-ordered deadlines.[10]

General Options for Congress

Most congressional committees conduct oversight hearings on agency activities and programs throughout the legislative session. Given the interest in the issues described in this report, it is likely that many oversight hearings will be held in the 112th Congress. If Congress decides to explore the way federal agencies regulate environmental issues, there are at least four sets of options available.

One option is the Congressional Review Act (CRA).[11] The CRA establishes special congressional procedures for disapproving a broad range of regulatory rules issued by federal agencies. Before any rule covered by CRA can take effect, the federal agency that promulgates the rule must submit it to Congress. If Congress passes a joint resolution disapproving the rule under procedures provided by the act, and the resolution becomes law, the rule cannot take effect or continue in effect. Also, the agency may not reissue either that rule or any substantially similar one, except under authority of a subsequently enacted law. The path to enactment of such a resolution could be a steep one and still subject to presidential veto. Overriding a veto requires a two-thirds majority in both the House and Senate.[12]

Another more comprehensive option would be to amend current law to modify the regulating agency's authority. Bills using this approach in connection with some environmental regulatory issues were introduced in the 111th Congress but were not enacted.[13] While this might be the preferred option by some, including the Administration, the challenges associated with crafting the specifics of a bill acceptable to a majority could remain difficult.[14]

From an agricultural perspective, this option may be even more challenging. While committees may exert certain oversight powers there are jurisdictional issues to be considered.[15] In many cases environmental laws with potential to affect agriculture originate outside of the House and Senate Agriculture Committees. Though the issues associated with agriculture could still be of interest within other committees, it might not be a central focus. On the other hand, agricultural interests in Congress have achieved some previous success on cross-jurisdictional issues.[16]

To provide a more detailed response to the issue than what might be permitted under the CRA, a third option would be to introduce freestanding legislation. By specifically identifying issues and prescribing regulatory direction, standalone legislation may address many of the issues with the current regulatory approach but still face challenges similar to those of amending existing law. While freestanding legislation could also amend existing law, this option may be designed to be more acceptable to Members willing to consider a delay of regulatory action, as opposed to overturning or blocking regulatory action altogether.[17] In effect, freestanding legislation could buy time for additional action to be taken by Congress.

Another option that Congress could consider is to include an amendment on the agency's appropriation bill that prevents funds from being used for specific actions. This was most recently done in the FY2010 Interior, Environment and Related Agencies Appropriations[18] in which provisions restricted or prohibited the use of EPA funds for certain climate change regulatory activities affecting livestock operations. Because neither the House nor the Senate Appropriations Committees reported FY2011 EPA appropriations in the 111th Congress and these provisions from FY2010 appropriations were carried forward, the restricting provisions remain in effect under the current continuing resolution, which expires March 4, 2011.[19]

Report Organization

The remainder of this report is organized under four broad subheadings: Air, Water, Energy, and Chemicals. Each section includes selected regulatory actions and provides background information and statutory authority followed by the current status of the rule or regulatory action and issues identified or raised by the agricultural community regarding the regulatory action. Finally, each section identifies the appropriate CRS specialist for additional information, which is also contained in *Table 1* below.

Table 1. CRS Specialists on Environmental Issues

Issue Area	CRS Specialist	Contact Information
Clean Water Act	Claudia Copeland	ccopeland@crs.loc.gov, 7-7227
Endangered Species Act	Lynne Corn	lcorn@crs.loc.gov, 7-7267
Clean Air Act, particulate matter	Rob Esworthy	resworthy@crs.loc.gov, 7-7236
Clean Air Act	Jim McCarthy	Jmccarthy@crs.loc.gov, 7-7225
Spill prevention	Jonathan Ramseur	jramseur@crs.loc.gov, 7-7919
Toxic Substances Control Act, chemicals	Linda-Jo Schierow	lschierow@crs.loc.gov, 7-7279
Agriculture-based biofuels, ethanol	Randy Schnepf	rschnepf@crs.loc.gov, 7-4277
Voluntary agriculture conservation	Megan Stubbs	mstubbs@crs.loc.gov, 7-8707
Clean Air Act, mobile sources, biofuels	Brent Yacobucci	byacobucci@crs.loc.gov, 7-9662

AIR

Agricultural production practices from both livestock and crop operations generate a variety of substances that enter the atmosphere, potentially creating health and environmental issues. Agriculture's effect on air quality rose to national importance in the 1930s when the conversion of native grasslands to cropland caused severe dust storms known as the Dust Bowl. The federal response to this phenomenon created many of the conservation outreach and education programs that remain in place today.[20] While dust storms of this proportion no longer occur in the United States, issues associated with soil erosion, particulates and farm chemical emissions, and livestock odor are still of concern.

The following section covers five federal regulations relating to air, including:

- Mandatory reporting of greenhouse gases (GHG);
- GHG emissions tailoring rule and the "cow tax;"

- Reductions of emissions from gasoline/diesel powered stationary engines;
- National ambient air quality standards (particulate matter and ozone); and
- Emergency Planning and Community Right-to-Know Act (EPCRA) and Comprehensive Environmental Response, Compensation, and Liability Act (CERCLA) reporting requirements.

Mandatory Reporting of Greenhouse Gases (GHG)

EPA was required by the FY2008 Consolidated Appropriations Act[21] "to develop and publish a ... final rule not later than 18 months after the date of enactment of this Act, to require mandatory reporting of greenhouse gas (GHG) emissions above appropriate thresholds in all sectors of the economy of the United States."

On October 30, 2009, EPA promulgated the final Greenhouse Gas Reporting Rule.[22] The rule requires suppliers of fossil fuels or industrial gases, manufacturers of vehicles and engines, owners or operators of electric power plants, and other – mostly industrial – sources to report their emissions of GHG to EPA annually, beginning in 2011. Covered entities are required to report to EPA if they emit 25,000 tons or more of carbon dioxide or the equivalent amount of five other GHG (methane, nitrous oxide, hydrofluorocarbons, perfluorocarbons, and sulfur hexafluoride and other fluorinated gases). About 10,000 facilities in 31 categories of sources were covered by the rule, as promulgated. EPA subsequently added 11 other categories of sources.

Status

The only agricultural sources covered by the Reporting Rule are manure management systems that emit methane and nitrous oxide in amounts greater than the reporting threshold. EPA identified six specific categories of agricultural sources that could be subject to the rule: beef cattle feedlots; dairy cattle and milk production facilities; hog and pig farms; chicken egg production facilities; turkey production; and broilers and other meat type chicken production. In all, EPA estimates that 107 livestock facilities nationwide would need to report under the rule.

In EPA's FY2010 appropriations act,[23] however, Congress included language barring EPA from using funds under that act to implement mandatory GHG reporting by manure management facilities. This prohibition was carried over into FY2011 by the continuing resolutions that have funded EPA's continued operation (currently P.L. 111-322). Therefore, despite the inclusion of manure management systems among the regulated entities, no agricultural sources are currently required to comply with the Reporting Rule.

Issues

For the facilities required to report, the rule imposes little cost because it only requires monitoring and reporting, and the monitoring does not require direct measurement of emissions. EPA considered requiring direct measurement of GHG emissions from manure management systems, but rejected the approach due to what it termed "the extreme expense and complexity of such a measurement program."[24] Instead, the agency promulgated an approach that allows the use of default factors, such as a system emission factor, for certain elements of the calculation, combined with the use of site-specific data (e.g., number of livestock). EPA estimated the total annual cost of the rule for the 107 potentially affected manure management facilities to be $300,000.

In comments on the proposed rule, a number of agricultural stakeholders noted that agriculture as a whole is responsible for less than 1% of total GHGs emitted and questioned why manure management systems in particular were included in the proposal. Other categories of agricultural sources, such as livestock enteric fermentation and soil management, emit larger amounts of methane and nitrous oxide. EPA explained that it did not include reporting by the other agriculture categories because, for those sources, no direct GHG emission measurement methods are available except for expensive and complex equipment. Using emissions estimates for such sources, instead of direct measurement, would have a high degree of uncertainty and could burden a large number of small emitters.

Commenters also expressed concern about the difficulty that livestock facilities might have in determining whether or not they are subject to the rule. In response to these comments, EPA modified the proposal to remove manure sampling requirements and instead will allow facilities to use default values for estimating emissions. The threshold table within the final rule (*Table 2*) identifies animal population threshold levels below which facilities are not required to report emissions.

Table 2. EPA Animal Population Threshold Below Which Facilities Are Not Required to Report GHG Emissions

Animal Group	Average Animal Population (Head)
Beef	29,300
Dairy	3,200
Swine	34,100
Poultry:	
Layers	723,600
Broilers	38,160,000
Turkeys	7,710,000

Source: U.S. Environmental Protection Agency, "Mandatory Reporting of Greenhouse Gases," 74 Federal Register 56485, October 30, 2009.

Notes: For all animal groups except dairy, the average annual animal population represents the total number of animals present at the facility. For dairy facilities, the average annual animal population represents the number of mature dairy cows present at the facility. For additional information, see Table JJ-1of the Environmental Protection Agency, "Mandatory Reporting of Greenhouse Gases," 74 Federal Register 56485, October 30, 2009.

GHG Emissions Tailoring Rule and the "Cow Tax"

EPA promulgated standards for GHG emissions from new light duty motor vehicles on May 7, 2010 (see "Motor Vehicle and Heavy-Duty Truck GHG Rule and Corporate Average Fuel Economy (CAFE) Standards" below).[25] The standards are not considered particularly controversial in and of themselves, but their implementation, on January 2, 2011, triggered two other requirements of the Clean Air Act (CAA) that apply to stationary sources. The first of these is a requirement that stationary sources emitting any air pollutant "subject to regulation" under the act must obtain a permit under Title V of the CAA (Title V permit) if they emit more than 100 tons per year of the pollutant subject to regulation. Agricultural sources, such as confined animal feeding operations (CAFOs), are among those that could potentially be subject to this permit requirement. Because permit applicants must pay a fee to cover the costs of administering the permit program, many in the agriculture community have referred to this requirement as the "cow tax."

The second requirement triggered by implementation of the motor vehicle standards is a requirement that new or modified stationary sources emitting more than 100 or 250 tons annually of any pollutant subject to regulation under the act must obtain pre-construction permits (referred to as "PSD" permits) and install Best Available Control Technology (BACT) to reduce emissions.

Status

On June 3, 2010, EPA promulgated a rule that sets high thresholds for the Title V permit and PSD/BACT requirements that would apply to GHG emissions.[26] EPA says that under the promulgated rule, the agency has not identified any agricultural sources that would be required to obtain permits for GHG emissions, and therefore none would be subject to BACT requirements.[27]

Under the rule, called the GHG "Tailoring Rule," the threshold initially is annual emissions of 75,000 tons of carbon dioxide equivalents, not 100 or 250 tons as required for other pollutants by the PSD and Title V permits. With this threshold, the nation's largest GHG emitters, including power plants, refineries, cement production facilities and about two dozen other categories of sources (an estimated 13,000 facilities in all, or nearly 70% of the nation's largest stationary source GHG emitters) are the only sources required to obtain permits. Farms, smaller businesses, and large residential structures (about 6 million sources in all these categories), which would otherwise be required to obtain permits after GHGs became subject to regulation, are shielded from permitting requirements, including permit fees.

The June 2010 Tailoring Rule does not permanently exempt smaller sources. EPA expects to lower the threshold, but not below 50,000 tons of GHG emissions, through separate rule-making that would take effect in 2013. Further, EPA and state permitting authorities, within five years of the rule's promulgation, would conduct a study of the permitting authorities' ability to administer more inclusive PSD and Title V permit programs. Within a year of the study's completion, EPA and state permitting authorities would conduct rulemaking for this phase of the program. The study might confirm the threshold, revise it, or establish other streamlining techniques for subsequent permitting activity. It is unclear how agricultural sources might be affected by rule changes in 2013 or thereafter.

In the FY2010 appropriations act for EPA,[28] Congress included a provision prohibiting EPA from using funds under the act to promulgate or implement any rule requiring the issuance of CAA Title V permits for GHG emissions associated with livestock production. This prohibition was carried over into FY2011 by the continuing resolutions that have funded EPA's continued operation.

Issues

The issues related to the Tailoring Rule are similar to those raised by the "Mandatory Reporting of Greenhouse Gases (GHG)" discussed above. The rule itself appears to exempt all agricultural sources by its high thresholds and the exclusion of fugitive emissions, but many are concerned with whether EPA intends to consider any agricultural sources as subject to regulation under future Clean Air Act GHG rules.

Reduction of Emissions from Gasoline/Diesel Powered Stationary Engines

On June 15, 2004, EPA promulgated emission control standards for hazardous air pollutants emitted by gasoline- and diesel-powered stationary engines. This is primarily of concern to agricultural operations that rely on gas and diesel engines for irrigation pumping. The standards are generally referred to as the RICE (Reciprocating Internal Combustion Engine) rules. Besides setting emission standards, the rules would have exempted these engines from emission controls during startup, shutdown, and periods of malfunction. On December 18, 2008, the D.C. Circuit Court of Appeals ruled that the standards must address emissions during all phases of operation, including periods of startup, shutdown, and malfunction. As a result, the court vacated and remanded the rules to EPA.

Status

EPA subsequently divided the standards into two regulatory actions. On March 3, 2010, the agency issued a final rule for existing diesel-powered stationary engines.[29] The rule will apply to more than 900,000 stationary engines that are used as generators and to power pumps in industrial and agricultural settings. EPA issued final emissions standards for existing stationary engines that burn gasoline, natural gas and landfill gas, known as spark ignition engines, on August 20, 2010.[30]

Issues

The proposed rules were criticized by some state permitting authorities and industry groups as being unworkable, difficult to enforce, and perhaps unnecessary in rural settings. In response to these comments, EPA stated that most engines used by agricultural sources are smaller than 300 horsepower, and will be subject only to required management practices (e.g., frequency of oil changes). Catalysts or other control equipment would not be required.

National Ambient Air Quality Standards (NAAQS) - Particulate Matter

National Ambient Air Quality Standards (NAAQS) are standards for outdoor (ambient) air that are intended to protect public health and welfare from harmful concentrations of pollution. NAAQS are at the core of the Clean Air Act, even though they do not directly regulate emissions. In essence, they are standards that define what EPA considers to be clean air. Once a NAAQS has been set, the agency, using monitoring data and other information submitted by the states, identifies areas that exceed the standard and must, therefore, reduce pollutant concentrations to achieve it. After these "nonattainment" areas are identified, state and local governments have up to three years to produce State Implementation Plans which outline the measures they will implement to reduce the pollution levels and attain the standards.

NAAQS have been set for six pollutants. The two that affect the largest number of areas are those for ozone and particulate matter (PM). Because some farming and livestock practices contribute to particulate matter emissions (e.g., dust) and because particulate matter and ozone can affect agricultural productivity, the agricultural community continues to show particular interest in these standards.

On October 17, 2006, EPA published its final revisions to the NAAQS for particulate matter (particulates, or PM) and the national air quality monitoring requirements.[31] EPA revised the primary PM NAAQS by strengthening the preexisting (1997) standard for "fine" particulate matter – PM2.5 – by lowering the allowable *daily* concentration averaged over 24-hour periods of PM2.5 in the air. The *annual* $_{PM2.5}$ standard was unchanged from the 1997 standard. The 2006 PM NAAQS also retained the 24-hour standard and revoked the annual standard for slightly larger, but still inhalable, particles less than or equal to 10 micrometers (PM10).[32]

On November 13, 2009, EPA published its final designations for the 2006 $_{PM2.5}$ NAAQS that include 120 counties and portions of counties in 18 states as nonattainment areas based on 2006 through 2008 air quality monitoring data. The majority of the roughly 3,000 counties throughout the United States (including tribal lands) were designated attainment/unclassifiable. States have three years from the effective date to submit State Implementation Plans

(SIPs), which identify specific regulations and emission control requirements that would bring an area into compliance with the standard. The EPA is not requiring new nonattainment designations for PM_{10} since the standards were not strengthened. The counties designated nonattainment for the PM_{10} NAAQS are primarily located in California, Arizona and Utah (discussed in the "Issues" section below).

Status

EPA has initiated the next round of the periodic review of the particulates NAAQS, targeting June 2011 for proposing changes to the standards.[33] Final standards would not likely be promulgated before Spring 2012 based on previous EPA target dates.[34] Potential risk reduction estimates and initial EPA staff recommendations reported in July and August 2010 draft assessments, and subsequent reviews by the Clean Air Scientific Advisory Committee (CASAC), call into question the adequacy of protection afforded by the current PM standards (discussed further below). EPA continues implementation of the 2006 PM NAAQS.

Issues

The PM_{10} NAAQS are generally more of a concern within the agricultural community than the PM2.5 NAAQS.[35] Because the PM_{10} standard was not strengthened under the revised 2006 PM NAAQS, no new areas, including rural or non-urban areas, will be designated as nonattainment for PM_{10}. The majority of the 3,000 counties throughout the United States (including tribal lands) were designated attainment/unclassifiable PM_{10} NAAQS. As of January 25, 2011, 42 of the original 87 PM_{10} designated nonattainment areas have been redesignated to attainment.[36] Primarily counties in California and the other western states have not yet come into attainment.

EPA draft assessments[37] in support of its next round of review of the particulates NAAQS have further stimulated considerable debate and comment, particularly within the agricultural community, with regard to further tightening the PM_{10} standards. With regard to the adequacy of the current primary 24-hour (daily) PM_{10} standard, EPA staff concluded that consideration should be given to retaining or revising the current standard downward in a June 2010 second external review draft policy assessment. In its August 2010 review of the draft assessment, CASAC included recommendations that "...the primary standard be revised downwards and not retained," and did not agree that the available scientific evidence strongly supports the upper bound standard level proposed by EPA staff,[38] instead favoring a more stringent level to "provide enhanced protection."

Source: Provided to CRS by the EPA Office of Air Quality Planning and Standards, January 25, 2011.

Notes: Areas not highlighted on the map are designated attainment/unclassifiable. There are no PM10 nonattainment areas in Hawaii, which was not included on the EPA map. For more information, see of CRS Report RL33254, Air Quality: EPA's 2006 Changes to the Particulate Matter (PM) Standards.

Figure 1. Status of PM10 Nonattainment Areas (based on 2006-2008 air quality).

In July and August 2010 letters to the EPA Administrator, some members of Congress raised concerns about EPA's July 2010 staff recommendations and the potential impacts on agriculture associated with tightening the PM10 standards, encouraging EPA to retain the current primary standards.[39]

National Ambient Air Quality Standards (NAAQS)—Ozone [40]

Under the CAA, EPA is to review the science for each of these standards every five years, and either reaffirm or revise the standard. The EPA

Administrator completed a review of the ozone NAAQS in March 2008, and made both the primary (health-based) and secondary (welfare-based) standards more stringent, but he did not set the standards within the ranges recommended by the independent panel of scientists that advises him (i.e., CASAC). He also rejected their advice to change the form of the secondary standard to better measure whether ozone concentrations were above levels needed to protect crops and forests from damage.[41] Challenged in court, EPA agreed to reconsider the March 2008 decisions (court decisions are discussed further below).

Status

On January 19, 2010, EPA proposed to strengthen the primary ozone NAAQS and to revise the form of the secondary standard as the agency's scientific advisers had recommended. Under the proposed revisions, the vast majority of counties with ozone monitors would be found in nonattainment of the primary standard, using the most recent available data, and many might violate the secondary standard, as well.

EPA expects to promulgate a final version by July 29, 2011. Between now and July 29, the agency plans to ask its Clean Air Scientific Advisory Committee (CASAC) to conduct a further review and allow for additional public comment. EPA is also proposing new monitoring requirements for the states, with more monitors to be placed in rural areas.[42]

Issues

After promulgation of a revised NAAQS, implementation takes several years. First, areas need to be designated "attainment" or "nonattainment." States then have three years to develop implementation plans identifying control measures. In the meantime, air quality is likely to improve as a result of regulations currently being phased in for cars, trucks, and electric power plants, among other sources.

Ultimately, however, the ozone NAAQS revision could be one of the more significant regulations promulgated by EPA, and could call attention to air quality problems in agricultural areas to a far greater extent than previous standards.

EPCRA and CERCLA Reporting Requirements

The Emergency Planning and Community Right-to-Know Act (EPCRA) and the Comprehensive Environmental Response, Compensation, and Liability Act (CERCLA, or Superfund) have reporting requirements that are triggered when specified quantities of certain substances are released to the environment, including ammonia and hydrogen sulfide. Both ammonia and hydrogen sulfide are chemicals generated by livestock manure, particularly swine and poultry, when in concentrated animal populations. Both CERCLA and EPCRA include citizen suit provisions that have been successfully used to take legal action against poultry and swine operations for violations of the reporting requirements of the laws. In 2005, a group of poultry producers petitioned EPA for an exemption from EPCRA and CERCLA release reporting requirements, arguing that releases from poultry growing operations pose little or no risk to public health, while reporting imposes an undue burden on producers and government responders.[43]

Status

In December 2008, EPA promulgated a EPCRA/CERCLA administrative reporting exemption for air releases.[44] The final rule exempts hazardous substance releases that are emitted to the air from all livestock operations (not just poultry farms) from CERCLA's requirement to report releases to the air to federal officials. It provides a partial exemption for such releases from EPCRA's requirement to report releases to state and local emergency officials: the final rule continues to apply EPCRA's reporting requirement to large CAFOs (those subject to Clean Water Act permitting, discussed below in the "Implementation of Existing Clean Water Act Permit Requirements for CAFOs" section), but exempts smaller facilities. The reporting exemptions in the final rule took effect January 20, 2009.

The 2008 rule was challenged by industry groups, including the National Pork Producers Council, as well as environmental advocates. Industry sought legal action, arguing that CAFOs should be exempted from all reporting under Superfund and EPCRA because air emissions from animal feeding operations pose no threat to public health or the environment. Environmentalists also went to court, arguing that CAFOs should report under both laws because air emissions from animal feedings operations do pose a public health and environmental risk. The legal challenges were consolidated in the U.S. Court of Appeals for the District of Columbia (*Waterkeeper Alliance v. EPA*, D.C. Cir., No. 09-1017). In June 2010 the government asked the court to remand the

2008 rule for reconsideration and possible modification. The court approved the government's request in October 2010, and although the rule remains in effect the future schedule is currently unknown.

Issues

The agriculture industry remains concerned about the potential burden on large CAFOs of complying with the EPCRA reporting requirements, even though the final rule exempted facilities that are not subject to Clean Water Act permitting (see "Implementation of Existing Clean Water Act Permit Requirements for CAFOs" below). Critics of the 2008 rule, including environmentalists and some state air quality officials, contend that the CERCLA and EPCRA reports provide good information about emissions that enable citizens to hold companies accountable in terms of how toxic chemicals are managed.

WATER

The release of sediment, nutrients, pathogens, and pesticides from agricultural production can degrade the quality of water resources. While it is widely believed that agriculture can have a significant impact on water quality, there is no comprehensive national study of agriculture's effect on water quality.[45] Several water quality assessments document degradation from agriculture practices; however, the extent and magnitude is difficult to measure because of its nonpoint nature.[46] Federal water laws, such as the Clean Water Act (CWA), largely do not regulate agricultural actors in many cases giving the regulatory responsibilities to the states. Constraints on agricultural production to reduce pollution discharges typically arise at the state level in response to local concerns.[47]

The following section covers four regulations relating to water, including:

- Implementation of existing Clean Water Act permit requirements for CAFOs;
- Chesapeake Bay protection and restoration;
- Florida nutrient water quality standards; and
- Spill prevention control and countermeasure (SPCC) plans.

Implementation of Existing Clean Water Act Permit Requirements for CAFOs

Under the CWA, while most of agriculture is exempt from federal regulation, large CAFOs are defined as point sources and thus are subject to the act's prohibition against discharging pollutants into U.S. waters without a permit. In October 2008, EPA issued a regulation to revise a 2003 CWA rule governing waste discharges from CAFOs. This action was necessitated by a 2005 federal court decision (*Waterkeeper Alliance et al. v. EPA*, 399 F.3d 486 (2nd Cir. 2005)), resulting from challenges brought by agriculture industry groups and environmental advocacy groups that vacated parts of the 2003 rule and remanded other parts to EPA for clarification.[48] The 2008 rule details requirements for permits, annual reports, and development of plans for handling manure and wastewater. Parts of the rule are intended to control land application of manure and wastewater.

Status

According to EPA, the 2008 rule applies to about 15,300 CAFOs that will need permit coverage (74% of the 20,700 CAFOs operating in 2008).[49] Under the rule, CAFOs were to obtain permits and develop and implement nutrient management plans by February 27, 2009.

Further legal challenges followed promulgation of the 2008 revised rule. Agricultural industry groups (although generally satisfied with the rule) filed lawsuits in several federal appellate circuits. Environmental groups also brought a legal challenge to the rule. The various petitions were consolidated in the U.S. Court of Appeals for the 5th Circuit where the case is pending.[50] In addition, EPA officials have reportedly been meeting with environmental plaintiffs, discussing the possible settlement of portions of the litigation that could involve additional regulatory changes. In December 2009, the court agreed to a joint request from EPA and environmentalists to sever the activists' portion of the litigation. In settling with environmental plaintiffs, EPA agreed to issue guidance aimed at clarifying what CAFOs must do to comply with federal clean water regulations and to help CAFO owners determine whether they need permits; the guidance was issued May 28, 2010. EPA also agreed to propose a rule within one year to collect facility information from all CAFOs, such as number of types of animals, type and capacity of manure storage or treatment process, and quantity of manure generated annually by the CAFO, in order to provide a CAFO inventory and assist in implementing the 2008 rule. Agricultural industry groups continue to challenge the rule; oral argument in the litigation occurred October 5, 2010. The 2008 rule remains in effect.

Issues

In the Second Circuit challenge to the 2003 rule, the agricultural industry had challenged a provision that explicitly required all CAFOs to apply for a National Pollutant Discharge Elimination System (NPDES) permit, or to demonstrate that they have no potential to discharge. The court upheld their argument that the CWA only requires facilities that actually discharge to seek permit coverage. Industry groups continue to disagree with any presumption that CAFOs do discharge pollutants, thus they objected to EPA's attempts in the 2008 revised rule to get CAFOs to voluntarily seek permits and the specific addition of a permit requirement for those that "propose to discharge." According to this view, EPA may not lawfully establish permitting requirements based on speculation as to possible future CAFO discharges. In addition, a number of questions linger about implementation of the 2008 rule. For example, agricultural industry groups are concerned that EPA regions may be providing differing interpretations of a provision of the 2008 rule that allows farms to self-certify that they will not discharge, a finding that allows them to avoid having to apply for a permit and protects CAFOs from liability for not having a permit in the event of an accidental discharge. Agricultural industry groups also are concerned that EPA will initiate a new rulemaking that would include additional permit and pathogen control requirements.

Chesapeake Bay Protection and Restoration

Despite several decades of activity by governments, the private sector, and the general public, efforts to improve and protect the Chesapeake Bay (Bay) watershed have been insufficient to meet restoration goals. Although some specific indicators of Bay health have improved slightly or remained steady (such as blue crabs and underwater bay grasses), others remain at low levels of improvement, especially water quality. Overall, the Bay and its tributaries remain in poor health, with polluted water, reduced populations of fish and shellfish, and degraded habitat and resources. The primary pollutants causing impairments are nutrients (nitrogen and phosphorus) and sediment discharged from multiple urban, suburban, and rural sources around the Bay.

In May 2009, President Obama issued an executive order that declared the Bay a "national treasure" and charged the federal government with assuming a strong leadership role in restoring the Bay.[51] The executive order established a Federal Leadership Committee for the Chesapeake Bay to develop and implement a new strategy for protecting and restoring the Chesapeake region. The resulting strategy, which was released in May 2010, launched major specific environmental initiatives to establish new clean water regulations on stormwater discharges and pollution discharges from animal feedlots in the Bay watershed, put new agricultural conservation practices on farms in the region, and restore land and water habitat.[52]

According to EPA, agriculture represents the single largest source of nutrient and sediment pollution to the Bay, with about half of agriculture's pollutant load directly related to livestock waste. Agriculture covers about 25% of the Bay watershed, and is the largest intensively managed land use in the watershed. EPA believes that excess livestock waste, improperly applied fertilizers, and certain cropland tillage practices increase nutrient and sediment discharges to the Bay.

A central feature of the overall strategy for restoring the Bay is EPA's establishment of a Total Maximum Daily Load (TMDL). Section 303 of the CWA requires states to identify waters that are impaired by pollution, even after application of pollution controls. For those waters, states must establish a TMDL to ensure that water quality standards can be attained. A TMDL is essentially a pollution budget, a quantitative estimate of what it takes to achieve standards, setting the maximum amount of pollution that a waterbody can receive without violating standards. If a state fails to do this, EPA is required by the CWA to make its own TMDL determination for the state. Throughout the United States—including the Chesapeake Bay watershed—more than 20,000 waterways are known to be violating applicable water quality standards and to require a TMDL.[53] Lawsuits have been brought with the intention of pressuring EPA and states to develop TMDLs, including for the Chesapeake Bay because the waters of the Bay have been identified as being impaired, that is, not meeting applicable water quality standards. The Chesapeake Bay TMDL is the largest single TMDL developed to date. It addresses all segments of the Bay and its tidal tributaries that are impaired from discharges of nitrogen, phosphorus, and sediment. The goal is to have TMDL implementation measures in place by 2025 to assure attainment and maintenance of all applicable water quality standards. The TMDL allocates needed reductions of these pollutants to all jurisdictions in the 64,000 square mile watershed, not to individual segments of streams or waterbodies, as is more typical of other TMDLs prepared by states or EPA.[54]

As part of the TMDL development process, states are to prepare Watershed Implementation Plans (WIPs) identifying specific reductions and control measures to achieve needed pollutant reductions from point sources (i.e., industrial and municipal facilities and CAFOs) and nonpoint sources (i.e., farms and forests), as well as two-year milestones to implement the plans. EPA fully expects that states will meet commitments and milestones in the WIPs, but the agency also has identified a number of potential actions currently available to it if a state fails to do so, including expanding permit coverage to currently unregulated sources, requiring net improvement offsets, conditioning EPA grants, or increasing federal enforcement in the watershed.

Status

Under a consent decree resolving some of the litigation over the Chesapeake Bay, EPA was required to establish a TMDL no later than May 1, 2011. EPA issued the TMDL on December 29, 2010—ahead of its self-imposed December 31 deadline.[55]

Issues

Specific requirements that could apply to agricultural operations generally, or in particular segments of the watershed, are speculative until WIPs are developed by states. Nevertheless, EPA's TMDL plans and the overall federal Bay restoration strategy under the 2009 executive order are controversial with agricultural and other groups that are concerned about the likely mandatory nature of many of EPA's and states' upcoming actions. Agricultural interests are concerned that farm operations in the Bay watershed will be subject to more regulation than competitors in other states, putting their operations at a significant competitive disadvantage. Many of these groups have also been concerned that the underlying scientific data and modeling used by EPA to develop the TMDL does not fully reflect ongoing voluntary efforts by agriculture to reduce pollutant discharges. A lawsuit challenging EPA's authority to set pollution limits under the multistate TMDL was filed by the American Farm Bureau Federation on January 10, 2011.[56] On the other hand, environmental activists in particular are pleased that the federal government is now asserting a leadership role to restore the Bay and have supported legislation that would codify requirements for the Bay TMDL in the CWA, while authorizing grants and other assistance for implementing required measures. Companion bills to do so were introduced in the 111th Congress (S. 1816 and H.R. 3852). In June 2010, the Senate Environment and Public Works Committee approved an amended version of S. 1816. As reported, the bill generally sought to codify 2025 as a date-certain for implementing restoration actions throughout the Chesapeake Basin and would have made explicit backup authority for EPA to develop measures to restore the watershed, if states fail to do so. The legislation would have authorized to be appropriated significant financial resources, totaling $2.26 billion over five years, to assist in implementing programs, projects, and measures for restoration of the Chesapeake Basin watershed. The House Agriculture Committee also approved separate legislation (H.R. 5509) that would have authorized an expanded role for USDA in Bay restoration. No further action was taken on either measure before the 111th Congress adjourned.

Florida Nutrient Water Quality Standards

The CWA directs states to adopt water quality standards for their waters and authorizes EPA to promulgate new or revised standards if a state's actions fail to meet CWA requirements. Water quality standards consist of designated uses, criteria to protect the designated uses and an antidegradation statement. They serve as the framework for pollution control measures that are specified for individual sources by states.

Status

Because of severe water quality impairment of Florida waters by nutrients (nitrogen and phosphorus) from diverse sources including agriculture and livestock, municipal and industrial wastewater discharges, and urban stormwater runoff, EPA determined in 2009 that Florida's existing *narrative* water quality standards for nutrients must be revised in the form of *numeric* criteria that will enable Florida to better control nutrient pollution. In 2009 EPA entered into a consent decree with environmental litigants requiring the agency to promulgate numeric nutrient water quality standards for Florida. To meet the legal deadline, EPA issued the first phase of these standards on November 15, 2010, establishing standards for lakes and flowing waters in the state. The EPA rule does not establish any requirements directly applicable to regulated entities or other sources of nutrient pollution. Water quality standards do not have the force of law until the state translates them into permit limits or otherwise imposes pollution control requirements on dischargers in the state.[57] In response to criticism of the proposed standards, EPA delayed the effective date of the final rule for 15 months, to allow local governments, businesses, and the state of Florida time to review the standards and develop implementation strategies. Nevertheless, separate legal challenges to the rule have been filed in federal court by environmental advocates, several industry groups, and the state of Florida's agriculture commissioner.

The second phase of standards is due to be proposed by November 14, 2011, and finalized by August 15, 2012. They will apply to estuarine, coastal waters, and Southern Florida inland flowing waters.

Issues

While few dispute the need to reduce nutrients in Florida's waters, EPA's proposal has been controversial, involving disputes about the data underlying the proposal, potential costs of complying with numeric standards when they are incorporated into discharge permit limitations, and disputes over administrative flexibility. EPA responds that the rule is intended to ensure the health of Florida's waterways and its economy, because the types of water quality problems associated with nutrients—algae blooms that are toxic to humans, fish and animals—have economic impacts throughout the state. Some groups also fear that EPA's actions in Florida, which was the first time that EPA has established statewide numeric nutrient standards, will be a precedent for similar regulatory action elsewhere. For example, environmental advocacy groups have petitioned or filed lawsuits seeking to require EPA to establish numeric nutrient water quality standards in Kansas and for the Upper Mississippi River Basin.

Spill Prevention, Control, and Countermeasure (SPCC) Plans

The Clean Water Act mandated regulations to prevent the discharge of oil from various sources.[58] Pursuant to this statutory requirement,[59] EPA crafted regulations for non-transportation-related facilities in 1973. Affected facilities must prepare and implement spill prevention control and countermeasure (SPCC) plans.[60] For example, SPCC regulations require secondary containment (e.g., dikes or berms) for certain oil-storage units; and plans must be certified by a Professional Engineer unless a facility owner/operator is able to self-certify the plan.

The EPA SPCC plan requirements apply to non-transportation-related facilities that drill, produce, store, process, refine, transfer, distribute, use, or consume oil or oil products;[61] and that could reasonably be expected to discharge oil to U.S. navigable waters or adjoining shorelines.[62] Facilities, including farms,[63] are subject to the rule if they meet at least one of the following capacity thresholds: an aboveground aggregate oil storage capacity greater than 1,320 U.S. gallons,[64] or a completely buried oil storage capacity greater than 42,000 U.S. gallons.

Status

Following the passage of the Oil Pollution Act of 1990,[65] EPA proposed substantial changes and clarifications to the SPCC regulations that were made final in July 2002.[66] EPA has both extended the 2002 rule's compliance date (on multiple occasions) and made further amendments to the 2002 rule.[67] Pursuant to a rule issued on October 14, 2010,[68] the current deadline for complying with SPCC requirements for most facilities is November 10, 2011.

Note that the July 2002 final rule and subsequent amendments did not alter the requirement for owners or operators of facilities in operation before August 16, 2002—the effective date of the 2002 final rule—to maintain and continue implementing their SPCC plans in accordance with the SPCC regulations in effect before the 2002 rulemaking.

Issues

Some of the recent SPCC rulemakings have included provisions that would affect agricultural operations. One issue that has received recent interest is the applicability of the SPCC requirements to milk containers.[69] Although milk was not specifically mentioned in the 2002 SPCC rulemaking, EPA subsequently clarified that certain milk containers would be subject to SPCC provisions, because milk would meet the SPCC regulatory definition of oil.[70] In 2009, EPA proposed a conditional exemption for milk storage units.[71] This exemption has not been finalized, but in its October 14, 2010 rule (which extended SPCC compliance for one year for most facilities) EPA provided a specific extension for facilities subject to milk storage SPCC provisions, providing an additional year extension to any final rule's compliance date that would apply to milk units.

In some cases, EPA appears to have taken different approaches to farms over time. For example, in a December 2006 final rule, EPA decided to extend the SPCC plan compliance date for small farms (i.e., total oil storage capacity of 10,000 gallons or less) "either indefinitely or until the Agency publishes a final rule in the *Federal Register* establishing a new compliance date." [72] EPA removed this provision in a June 2009 final rule, establishing the same compliance dates for farms as for all other facilities. In addition, in its December 5, 2008 rulemaking,[73] EPA specifically excluded farms from the loading/unloading rack requirements.[74] However, in its November 2009 final rule, EPA removed this exclusion, concluding that "certain facilities (i.e., farms and oil production facilities) should not be treated differently than other facilities, even if loading/unloading racks are not typically associated with these types of facilities."[75]

However, the most recent substantive rulemaking in November 2009, included some amendments that may benefit farming operations. The rule exempts pesticide application equipment and related mix containers that may currently be subject to the SPCC rule when crop oil or adjuvant oil are added to formulations. EPA also clarifies that a nurse tank is considered a mobile refueler, and, like other types of mobile refuelers, is exempt from the sized secondary containment requirements. EPA estimated that the total cost savings to farm owners and operators from these (and other) amendments are estimated at $13 million on an annualized basis (2007$).[76]

ENERGY

The agricultural industry is sensitive to fluctuations in energy sources and cost. The use of petroleum-based fertilizers, diesel fuel, and, more recently, corn-based ethanol all have a significant impact on both crop and livestock operations. Since the 1970s, federal policies have offered a variety of incentives, regulations, and programs to encourage growth in the bioenergy industry as a sustainable alternative to fossil fuels.[77] The increased emphasis on agriculture-based biofuels has received mixed reviews within the agricultural community.[78] While some continue to push for greater federal involvement, critics of the federal intervention also have emerged.

The following section covers four federal regulations relating to energy, including:

- Motor vehicle and heavy-duty truck greenhouse gas (GHG) rule and Corporate Average Fuel Economy (CAFE) standards;
- Renewable fuels standard (RFS2) rule; and
- E15 waiver petition.

Motor Vehicle and Heavy-Duty Truck GHG Rule and Corporate Average Fuel Economy (CAFE) Standards

The Energy Policy and Conservation Act of 1975 (EPCA)[79] requires car and light truck manufacturers to meet corporate average fuel economy (CAFE) standards. The Energy Independence and Security Act of 2007 (EISA)[80] requires the National Highway Traffic Safety Administration (NHTSA) to develop rules to tighten CAFE standards and to promulgate fuel economy standards for medium- and heavy-duty trucks, reflecting the "maximum feasible improvement" in fuel efficiency.

In response to a 2007 Supreme Court decision (*Massachusetts v. EPA*),[81] EPA is required to, among other things, determine whether GHGs from automobiles endanger public health and welfare. On December 7, 2009, EPA issued such an "Endangerment Finding." Thus, under the Clean Air Act (CAA), EPA is required to promulgate rules on emissions of GHGs from motor vehicles. Because fuel economy and vehicle GHG emissions are tightly linked, the Administration proposed light-duty vehicle regulations in September 2009 that would integrate fuel economy and GHG rules into one process;[82] regulations for model year (MY) 2012-MY2016 were finalized in May 2010[83] and in October 2010 EPA and NHTSA announced their intent to propose similar regulations for MY2017-2025.[84] On November 30, 2010, EPA and NHTSA proposed integrated GHG and fuel economy standards for medium-and heavy-duty vehicles.[85]

Status

On May 7, 2010, EPA and NHTSA finalized rules to integrate CAFE standards with light-duty vehicle GHG standards. The Administration estimates that the rule will reduce lifecycle costs for most vehicle purchasers, as fuel savings are expected to more than offset the increase in purchase price ($1,100). The new standards will be phased in beginning with MY2012. While the rulemaking process was combined, EPA and NHTSA have recognized that some parts of the GHG program will not translate to the CAFE program, and vice versa.[86]

In October 2010 EPA and NHTSA announced their intent to propose further regulations for MY2017-2025,[87] and on January 24, 2011, EPA announced that it and NHTSA would propose MY2017-2025 standards in the same time frame as the state of California – by September 1, 2011.[88]

EPA's endangerment finding specifically referenced medium- and heavy-duty trucks as among sources that contribute to GHG emissions. Proposed heavy-duty truck GHG and fuel economy standards would be phased in between 2014 and 2018. EPA estimates that the rule will reduce lifecycle vehicle costs, factoring in the fuel savings and increase in purchase price.[89] EPA estimates that because of fuel savings most truck owners would see a payback period of one to five years.[90] In their Rulemaking Gateway, EPA projects completion of the final rule by August, 2011.[91]

Issues

This issue has a somewhat indirect effect on agriculture. The fact that vehicle purchase prices are expected to increase may affect agricultural producers who purchase cars, light trucks, and heavy trucks for use in their businesses (including light-duty and super-duty pickups, vans, and flatbed trucks). While for most purchasers those increases will be offset by lower fuel expenditures over the lifetime of these vehicles, the increase in up-front costs may influence some agricultural producers' decisions to purchase new vehicles notwithstanding the expected lifecycle cost savings. The proposed heavy-duty rules do not directly apply to non-road engines and equipment, but because many heavy-duty diesel engines are used in both on-road and non-road applications (including farm equipment), some stakeholders are concerned that compliance with rules could raise the cost of diesel engines in general.

Renewable Fuels Standard (RFS2) Rule

The Energy Independce and Security Act (EISA) expanded the renewable fuel standard (RFS) established in the Energy Policy Act of 2005.[92] The RFS requires a significant growth in U.S. biofuel use. In 2011, the RFS mandate is 13.95 billion gallons of biofuels from various sources (consisting mostly of ethanol from corn starch). By 2022, EISA will require 36 billion gallons of biofuel be used in the nation's fuel supply. Within the larger RFS, EISA mandates the growing use of advanced biofuels (i.e., non-corn starch biofuels), including fuels produced from cellulosic feedstocks. By 2022, the advanced biofuels mandate grows to 21 billion gallons, including 16 billion gallons of cellulosic biofuel.[93]

EISA also requires that advanced biofuels—e.g., cellulosic biofuels, biomass-based diesel substitutes, and other advanced biofuels—as well as conventional biofuels from newly built refineries, meet certain lifecycle GHG reduction requirements.[94] EPA is required to classify biofuel production based on their lifecycle emissions, including emissions from direct and indirect changes in land use. Only fuels that achieve a 50% reduction in GHG emissions relative to petroleum fuels may be classified as advanced biofuels. Cellulosic biofuels must achieve at least a 60% GHG emission reduction, while fuels from new corn ethanol plants must achieve a 20% GHG emission reduction – corn ethanol plants in existence or under construction when EISA was enacted (December 19, 2007) are grandfathered.

Status

On February 3, 2010, EPA finalized new rules for the expanded renewable fuel standard (RFS2).[95] These rules were effective July 1, 2010, but covered biofuel production for all of 2010. In 2011, the RFS2 requires the use of 13.95 billion gallons of ethanol and other biofuels in transportation fuel. Within the larger mandate, the RFS2 requires the use of 1.35 billion gallons of advanced biofuels (fuels other than corn starch ethanol) in 2010, including 6.6 million gallons of cellulosic biofuels. Within the rules, EPA finalized procedures for fuel suppliers to generate credits under the system—credits that can be sold or traded. EPA also finalized methodologies for determining lifecycle GHG emissions.

Issues

The RFS has been a major policy supporting the development of U.S. biofuels industries, especially for corn-based ethanol producers. Many believe that the expanded RFS2 will be the main pillar of support for existing U.S. biodiesel production capacity (due to the uneconomical nature of U.S. biodiesel production). In future years, as the advanced biofuel mandates grow, the RFS could be the key driver for the development of biofuels from cellulose, algae, and other nonfood/feed commodities.

RFS expansion could lead to concomitant pressure on limited agricultural resources (most notably land) as feedstock production intensifies on existing cropland and expands onto new, marginal lands. This may raise the general price level for those commodities that compete for the affected cropland, as well as having important secondary effects in related agricultural markets including livestock feed markets and agricultural input markets. As a result, the potential for unintended consequences (e.g., land use, commodity prices) in non-biofuels markets could increase.

Expanding cultivation onto marginal lands (including reclaimed Conservation Reserve Program acres) and more intensive cultivation (including increased water, pesticide, and fertilizer use) on existing cropland is expected to put new pressures on environmental resources. This could also put substantial pressure on the agricultural research infrastructure to develop technologies or techniques that enhance per-acre productivity in an effort to mitigate unintended price pressures and secondary market effects.

The clearest example of increasing pressure on resources (with unintended consequences) is the rapid growth of corn use for ethanol production. During the 2005/06 crop year, corn ethanol production used 1.6 billion bushels of corn or about 14.4% of U.S. production. This usage share has grown in lockstep with the RFS mandate. In the current 2010/11 crop year, corn ethanol production is expected to approach 12 billion gallons while consuming over 4.9 billion bushels or nearly 40% of the 2011 corn harvest.[96] While U.S. corn production has expanded and is expected to continue to expand (primarily due to continued yield growth as corn area expansion is thought to be very near its sustainable maximum), corn use for ethanol has expanded even faster. As a result, corn prices have moved steadily higher. The 2005/06 crop year farm price for corn was $2.00 per bushel. The farm price of corn was $4.20 per bushel in 2007/08, $4.06 in 2008/09, and is projected at $5.30 in 2010/11.[97] Corn is the primary feed ingredient used by the U.S. livestock sector representing over 90% of all grains consumed, and about 57% of all grains and feed concentrates consumed annually. As the price of corn rises, the entire feed complex price structure has risen putting a cost squeeze on the U.S. livestock sector. In the long run, an intensification of this pressure could lead to regional shifts in comparative advantage in certain livestock production activities that could increasingly favor proximity to corn ethanol plants for access to the co-product distiller's dried grains and solubles.

As corn ethanol production grows to 15 billion gallons by 2015, it may continue to compete with other corn users and keep upward pressure on commodity prices. If the unintended consequences of RFS expansion are sufficiently large or dramatic as some have suggested,[98] policymakers may experience pressure to waive future RFS mandates.

Another key issue is the role of cellulosic biofuels in the RFS2. Cellulosic biofuels are in their infancy, and no commercial-scale refineries have begun operation as of early 2011.[99] Because of this, EPA had to use its waiver authority under EISA to reduce the mandated 2010 level for cellulosic biofuels from 100 million gallons to 6.5 million gallons (a decrease of over 90%). For 2011, EPA reduced the cellulosic mandate from 250 million gallons (as scheduled in EISA) down to 6.6 million gallons.100 If commercial capacity does not come online rapidly, EPA may need to issue another waiver in 2012 (the cellulosic mandate is scheduled at 500 million gallons). It is unclear what effect the delays in implementing the cellulosic biofuel mandate will have on investment and in the development of the cellulosic biofuel industry.

E15 Waiver Petition

By 2022, EISA requires the use of 36 billion gallons of renewable fuels, and much of this could be ethanol from a variety of feedstocks (many of which are agricultural-based, see "Renewable Fuels Standard (RFS2) Rule" discussion above). However, there is an obstacle to the use of this quantity of ethanol in gasoline. Currently, although some ethanol is sold as an alternative fuel (E85), most is sold as an additive in conventional and reformulated gasoline. Until recently, the amount of ethanol that could be blended into gasoline for all uses was limited to 10% by volume (E10) pursuant to EPA guidance under the CAA, as well as by vehicle and engine warranties, and certification procedures for fuel dispensing equipment.

As the RFS is structured, assuming that most of the mandate is met using ethanol, the volume of ethanol blended in gasoline is limited by gasoline consumption. In 2012, the RFS will require over 15 billion gallons of renewable fuel, while projected gasoline consumption in 2012 is slightly less than 150 billion gallons. After 2012, the renewable fuel mandate will continue to increase. However, a limit of 10% ethanol means that ethanol for gasoline blending (not including E85) likely cannot exceed 15 billion gallons per year.[101] This "blend wall" is the maximum possible volume of ethanol that can be blended into U.S. motor gasoline. The actual limit could be lower since older fuel tanks and pumps at some retail stations may not be equipped to handle ethanol-blended fuel.[102]

Status

On March 6, 2009, Growth Energy (on behalf of 52 U.S. ethanol producers) applied to EPA for a waiver from the CAA limitation on ethanol content in gasoline. Until recently, ethanol content in gasoline for all uses was capped at 10% (E10); the application requested an increase in the maximum concentration to 15% (E15). If fully granted, the waiver would allow the use of significantly more ethanol in gasoline than is currently permitted.

On November 4, 2010, EPA granted a partial waiver allowing the use of E15 in MY2007 and newer vehicles.[103] The agency delayed a decision on MY2001-2006 vehicles until the Department of Energy completed testing of those vehicles. On January 21, 2011, EPA announced that the waiver would be expanded to include MY2001-2006 vehicles.[104] EPA determined that data were insufficient to address concerns that had been raised over emissions from MY2000 and older vehicles, as well as heavy-duty vehicles, motorcycles, and non-road applications (including farm equipment), and thus a waiver for these vehicles/engines was denied. EPA has noted that granting the waiver eliminates only one impediment to the use of E15 – other factors, including retail and blending infrastructure (including gasoline storage tanks and pumps), state and local laws and regulations, and manufacturers' warranties, would still need to be addressed. Because of concerns over potential damage by E15 to equipment not designed for its use, this partial waiver has been challenged in court by a group of vehicle and engine manufacturers.[105]

Issues

EPA approval of the waiver request could help open the door to E15 blending. This could be a strong signal to the biofuels industry concerning federal support for meeting and enforcing RFS mandate levels. As a result, this could help to stimulate new investment in the biofuels sector. In the short run the corn ethanol industry would be the main beneficiary since it is best able to respond to the expanding RFS mandates. Any further increase in corn ethanol use would benefit corn producers. The net result would be an intensification of agricultural resource use with the same consequences discussed previously (see "Renewable Fuels Standard (RFS2) Rule").

The ability to address concerns over the use of E15 in legacy equipment (both infrastructure and vehicles) will affect the roll-out of E15 to retail stations. As noted above, EPA's decision to allow E15 in some vehicles only addresses one part of the blend wall. State laws and regulations, vehicle and equipment certifications and warranties, the questions over fuel suppliers willingness to market the fuel could all be impediments to an expansion of E15 use. The result of equipment manufacturers' legal challenge to the partial wavier will be a key factor.

CHEMICALS

Agricultural "pests" can interfere with the production of crops and livestock used for food and fiber. Pests may include insects, plant pathogens, weeds, and vertebrates. If in abundance, pests may affect crop yield and cause a decline in quality. Hundreds of chemical products are available to repel or kill pests that affect agricultural production. Each uses different active ingredients, has a different potency, and a different impact on human health and the environment. The federal regulation of these chemicals includes registering and restricting their use.

The following section covers four federal regulations relating to chemicals, including:

- Disclosure of pesticide inert ingredients;
- Clean Water Act permits for pesticide application;
- Pesticide drift labeling;
- Atrazine; and
- Endangered Species Act (ESA).

Disclosure of Pesticide Inert Ingredients

Pesticide products generally contain active ingredients that are intended to control targeted pests as well as inert ingredients which are included to dilute the active ingredients, increase their ability to penetrate or adhere to leaf surfaces, or otherwise aid in the distribution and effectiveness of the pesticide product. Inert ingredients are not "active ingredients," but they are not necessarily chemically inert. Some inerts are potentially toxic or otherwise hazardous.

The Federal Insecticide, Fungicide, and Rodenticide Act (FIFRA) directs EPA to regulate the sale and use of pesticide products and pesticide labels by establishing requirements for pesticide labels. Use of a pesticide in a manner that is inconsistent with label instructions is a violation of FIFRA. One requirement for pesticide labels is a listing of active ingredients. No listing is required for most inert ingredients, but labels must indicate the total percentage of the product that is inert.

EPA has received two petitions requesting disclosure of certain potentially hazardous inert ingredients on pesticide labels. One petition was from 22 non-governmental organizations, while the other was from the Attorneys General of 15 U.S. states and territories. The petitioners requested that EPA require disclosure of certain inert ingredients that have been designated as hazardous under other environmental statutes. In response to the petitions, EPA is considering regulatory and voluntary options for providing information to the public about the identities of inert ingredients in pesticide products. According to EPA, it has the authority to require disclosure if the Administrator "determines that such ingredient(s) may pose a hazard to man or the environment."[106] In 1987, EPA required disclosure on pesticide labels of the identities of approximately 50 "inerts of toxicological concern."[107] A future rulemaking might expand this disclosure requirement to hundreds of additional chemicals and mixtures.

Status

EPA issued an advanced notice of proposed rulemaking on December 23, 2009.[108] Comments on options for providing public information closed on April 23, 2010. In the fall 2010 Regulatory Agenda, EPA classified this rulemaking as a "Long Term Action," suggesting that action is not imminent. The agenda also indicated that a regulatory analysis will accompany any proposed rule.[109]

Issues

Pesticide manufacturers often claim the identities of inert ingredients to be proprietary, and disclose them only to EPA and its contractors under a "confidential business information" agreement. Sometimes even the registrants of pesticide formulations are not told the identities of proprietary ingredients or mixtures supplied by manufacturers. EPA and the petitioners believe that registrants and consumers should be able to ascertain whether the products they use contain potentially hazardous ingredients. With such information, many believe the market should operate more efficiently by allowing formulators and consumers to choose products that include or exclude such ingredients rather than rely on government regulators to determine what ingredients are safe. EPA has announced that it is committed to improving public availability of such information to assist consumers and users of pesticides in making informed decisions and to reduce the presence of potentially hazardous ingredients in pesticides. After EPA required disclosure of 50 inerts in 1987, most of them were removed from pesticide products.[110] On the other hand, pesticide manufacturers might object to disclosure if it would reveal information deemed to be proprietary, lead to loss of sales, or jeopardize market advantage relative to competitors. This issue could be of interest to the agriculture community given the use of pesticide products by producers.

Clean Water Act Permits for Pesticide Application

For the more than 30 years since Congress enacted the Clean Water Act (CWA) and Federal Insecticide, Fungicide, and Rodenticide Act (FIFRA), little apparent direct conflict existed between the two laws. EPA's operating principle during that time was that pesticides used according to the requirements of FIFRA do not require regulatory consideration under the CWA. EPA had never required CWA permits for use of FIFRA-approved materials, and EPA rules did not specifically address the issue.[111]

Recently, however, EPA's interpretation and operating practice were challenged in several court cases. At issue has been how FIFRA-approved pesticides that are sprayed over or into waters are regulated and, specifically, whether the FIFRA regulatory regime is sufficient alone to ensure protection of water quality or whether such pesticide application requires approval under a CWA permit. The issue arose initially over challenges to some routine practices in the West (weed control in irrigation ditches and spraying for silvicultural pest control on U.S. Forest Service lands). It drew more attention in connection with efforts by public health officials to combat mosquito-borne illnesses such as West Nile virus. The litigation created uncertainty over whether application of pesticides and herbicides to waterbodies requires a CWA water discharge permit.

Status

EPA tried to promulgate policy to clarify the relationship of the two laws and to address conflicts resulting from several judicial rulings, ultimately in a regulation issued in November 2006 that attempted to specify circumstances in which pesticides applied to U.S. waters do not require CWA permits. That rule was challenged by multiple parties, and in January 2009, a federal appellate court vacated the rule.[112] As a result, persons who spray pesticides on or near water are now required to obtain a CWA permit.

The federal court's ruling appeared to leave little room for EPA to fashion a new rule consistent with the agency's long-standing view that FIFRA-compliant applications do not require CWA permits. Industry groups subsequently petitioned the Supreme Court to review the case, but the Court denied the petition on February 22, 2010. Legislation intended to nullify the 2009 federal court ruling was introduced in the 111th Congress (H.R. 6087/S. 3735 and S. 6273), but no further legislative action occurred.

The federal government did not seek a rehearing of the case. Instead, the government petitioned the court for a two-year stay of the order, to give EPA time to work with states and the regulated community to develop a general permit for pesticide applications covered by the decision (in lieu of individual permits). The court granted EPA's request in June of 2009. As a result, the court's ruling is scheduled to take effect on April 9, 2011. EPA proposed a draft permit in compliance with the court's order June 4, 2010, and plans to finalize the permit soon.[113] When the general permit is issued, EPA estimates that the universe of affected activities that for the first time will be subject to CWA permits is approximately 5.6 million applications annually, which are performed by 365,000 applicators covering four use patterns: (1) mosquito and other flying insect pest control; (2) aquatic weed and algae control; (3) aquatic nuisance animal control; and (4) forest canopy pest control. The permit will cover about 500 different pesticide active ingredients that are contained in approximately 3,700 product labels.

The draft permit applies to a variety of entities, including agricultural interests involved in crop and timber tract production, forest nurseries, and operating irrigation systems; pesticide and agricultural chemical manufacturing; mosquito or other vector control districts and commercial applicators that service them; utilities (e.g., electric power, natural gas, water supply and wastewater); and government agencies and departments engaged in air and water resource management and conservation. It requires all operators to minimize pesticide discharges to waters by practices such as using the lowest effective amount of pesticide product that is optimal for controlling the target pest. It also requires operators to prepare pesticide discharge management plans to document their pest management practices. Permittees must monitor for observable adverse effects in the treatment area and where the pesticides are discharged to U.S. waters. The permit will not cover agricultural stormwater runoff or irrigation return flow, as these discharges are statutorily exempt from CWA permitting, and it also will not cover terrestrial application to control pests on agricultural crops or forest floors. The EPA general permit will apply in states and areas where EPA is the NPDES permitting authority, but it is expected to be a model for other states to develop their own general permits.[114]

Issues

General permits cover categories of point sources that have common elements and that discharge the same types of wastes. They allow the permitting authority to allocate resources efficiently, especially when there is a large number of potential permittees. Permitting procedures are streamlined and simplified, compared with CWA individual permits. Still, many agricultural industry groups are fearful that the court's ruling and EPA's general permit will lead to more burdensome and potentially costly requirements.

Pesticide Drift Labeling

State agencies and EPA receive numerous complaints every year claiming harm (or risk) to beneficial insects or to human health from exposure to pesticides that have drifted beyond the fields targeted for application. Current federal and state regulations aim to protect agricultural workers and non-target animals and plants, but opinions differ about the adequacy of such regulations. Drift issues were addressed in recent years by an EPA advisory committee of stakeholders which recommended revisions to pesticide product labels to improve clarity and consistency, making the label instructions more comprehensible for applicators and facilitating enforcement by states. EPA issued proposed guidance in response to these recommendations.[115]

Pesticide drift is also the subject of a citizen petition received by EPA on October 13, 2009. Earthjustice, Farmworker Justice, and several other organizations requested that EPA assess exposure of children to pesticide drift and incorporate this information into risk assessments in support of registration decisions. They also requested interim prohibitions on the use of certain pesticides near homes, schools, and other places where children congregate.

Status

EPA issued the proposed labeling guidance and requested public comments on the citizen petition on November 4, 2009. The original periods for public comment on the labeling proposal and the petition were extended, but the comment periods ended March 5, 2010.

Issues

Some public health advocacy groups argue that the proposed label changes are "too little, too late." Nevertheless, bee keepers and some state enforcement officials urge rapid adoption of the label changes. Thirty-eight congressional representatives signed a letter dated November 20, 2009, asking EPA to require no-spray buffer zones for drift-prone pesticides of at least 60 feet for ground applications and 300 feet for aerial applications from homes, schools, parks, day care centers, and other places where children may congregate.[116]

On the other hand, some producers, pesticide applicators, and agricultural groups argue that the proposed label language is too vague and would invite litigation. For more than a year, according to lawyers for pesticide producers,

> [EPA] has maintained its position that its policies will define unallowable drift to be that which "may cause" harm from the pesticide—which the users and registrants of pesticides believe to be an extreme (and not authorized) extension of the current FIFRA standard of "does not cause unreasonable risk." This distinction is more than semantics, since the criticism of EPA's position is that it would provide for a subjective standard placing user [sic] of pesticide who followed every label instruction in jeopardy of a possible enforcement action even if "harm" has not occurred. The debate has been going on for years, and EPA has attempted to reassure critics that no new, more restrictive, standard is being imposed.[117]

Atrazine

Atrazine, a herbicide in use for at least 50 years, is one of the most widely used agricultural pesticides in the United States today.[118] It is used primarily on corn and sorghum in the Midwest. Atrazine is particularly useful for controlling broadleaf and grassy weeds in fields where no-till or low-till methods are employed to reduce topsoil erosion. These and other uses of atrazine are licensed by EPA which registers pesticide active ingredients, as well as formulated products, for specified uses under specified conditions under the Federal Insecticide, Fungicide, and Rodenticide Act (FIFRA) and the Federal Food, Drug and Cosmetic Act (FFDCA). The latter law applies only to pesticides used on food and animal feed crops. For more information about pesticide laws, see CRS Report RL31921, *Pesticide Law: A Summary of the Statutes.*

Recommended rates of application and other conditions of atrazine use are specified on the EPA-approved labels of various formulated pesticide products. It is illegal to use any pesticide product in a manner inconsistent with label instructions.

Widespread and relatively heavy use of atrazine, its persistence in the environment, reports of atrazine contamination of surface and drinking water,[119] and scientific studies indicating that exposure to atrazine might disrupt the normal action of hormones in animals[120] have prompted EPA's pesticide office to review atrazine registration more frequently than it has reviewed most other pesticide registrations. Scrutiny of atrazine began at least 20 years ago, and has continued, as new scientific hypotheses and studies have developed. For example, in November 1994, EPA initiated a "Special Review" of the potential risks posed by atrazine and related triazine pesticides to agricultural workers and to drinking water consumers.[121] This review is ongoing in February 2011. EPA issued a re-registration eligibility decision (RED) for atrazine April 6, 2006.[122]

In recent years, many scientific studies have been published indicating possible risks posed by atrazine to animals, as well as to human health. In addition, new monitoring data collected in the Midwest by Syngenta, a major manufacturer of atrazine, recently became available for analysis. In response, EPA announced October 7, 2009,[123] that it would again re-evaluate atrazine research. On the basis of its review, EPA will decide whether further regulatory restrictions are necessary to prevent unreasonable effects on human health or the environment. EPA called for the assistance of its FIFRA Scientific Advisory Panel (SAP) to review the Agency's plans for evaluating human epidemiological studies as well as studies of laboratory animals and wildlife.

Status

On November 3, 2009, EPA presented its plan for the atrazine reevaluation to the SAP.[124]. In 2010, the Agency held three SAP meetings to address atrazine issues. In 2011, another peer review is planned concerning the findings of the Agricultural Health Study, a large epidemiological study of agricultural workers and their families. That study is evaluating the potential association between human atrazine exposure and cancer risk.[125] According to EPA, "The SAP's recommendations will help EPA determine the appropriate next steps in the Special Review regarding cancer and drinking water issues."[126] The conclusions of the SAP also might influence EPA's risk assessment of atrazine and subsequent reregistration decisions.

Issues

Some policymakers and industry leaders are concerned about the continuing reviews of atrazine and similar herbicides. Chemical producers, distributors, and users are concerned that these reviews may lead to new restrictions or cancellation of pesticide uses. The potential cost to growers and consumers if EPA would cancel or restrict registration for atrazine could be considerable. On the other hand, public health advocates, some consumers of drinking water, and advocates for environmental protection, have argued that new restrictions on atrazine uses should be considered and may be warranted if current regulations do not ensure with a reasonable certainty that atrazine use on food will pose no harm to human health and that atrazine use in general will not pose an unreasonable risk to the environment.

Endangered Species Act (ESA)

The Endangered Species Act (ESA)[127] protects species identified as endangered or threatened with extinction and attempts to protect the habitat on which they depend. It is administered primarily by the Fish and Wildlife Service (FWS), and by the National Marine Fisheries Service (NMFS) for certain marine and anadromous species. Dwindling species are listed as either endangered or threatened according to assessments of the risk of their extinction. Once a species is listed, legal tools are available to aid its recovery and to protect its habitat. The ESA can become the visible focal point for underlying situations involving the allocation of scarce or diminishing lands or resources, especially in instances where societal values may be changing, such as for the forests of the Pacific Northwest, the waters of the Klamath River Basin, or the polar environment.

Status

In the case of agriculture, actions of some federal agencies may affect a very wide area or a region and have the potential to affect many listed species. Perhaps the most widely known of such agency actions is the registration and use of pesticides, such as those described in the "Pesticide Drift Labeling" section above. Where a substance can flow or be blown well outside its area of use, EPA would need to consult on registration of the new pesticide, and on any restrictions on its use (such as total area, weather conditions, distance from a particular habitat type, etc.). From December 2008, to May 4, 2009, somewhat broader regulations were in effect that were intended, among other things, to "narrow the circumstances when Federal agencies are required to consult with the Services [FWS or NMFS]."[128] The temporary change might have allowed agencies such as EPA to carry out formal consultation internally; the change, while offering the potential for speed or streamlining, was criticized as likely to create internal conflicts of interest within the action agencies.[129]

Issues

For activities on privately owned land such as farms and ranches, the primary direct impact of the ESA is through the law's prohibitions on taking of listed species. The word *take* means "to harass, harm, pursue, hunt, shoot, wound, kill, trap, capture, or collect, or to attempt to engage in any such conduct."[130] Thus, such activities as cutting down a tree that contains the nestlings of an endangered bird would constitute a taking. Plants have substantially less protection under the ESA, so removing an endangered plant on private land would trigger an ESA violation only under extremely limited circumstances.[131]

If federal actions (or actions of non-federal parties that require a federal approval, permit, or funding) might adversely affect a listed species as determined by FWS (or NMFS, depending on the species), the federal action agencies must complete a biological assessment.[132] The assessment is used to determine whether formal consultation is necessary. [133] Through consultation with either FWS or NMFS, federal agencies must ensure that their actions are "not likely to jeopardize the continued existence" of any endangered or threatened species, nor to adversely modify critical habitat.[134] This is referred to as a Section 7 consultation. "Action" includes any activity authorized, funded, or carried out by a federal agency, including permits and licenses.

End Notes

[1] A total of 1.9 billion acres of land and water cover the contiguous 48 states, of which 71% is non-federal rural land (nearly 1.4 billion acres). Non-federal rural lands are predominantly rangeland (409 million acres), forest land (406 million acres), and cropland (357 million acres). Source: USDA, NRCS, *2007 National Resources Inventory*, Summary Report, Washington, DC, December 2009, p. 6, http://www.nrcs.usda.gov/technical/NRI/2007/2007_NRI_Summary.pdf .

[2] For more information, see CRS Report R40763, *Agricultural Conservation: A Guide to Programs*.

[3] Michelle Perez, Craig Cox, and Ken Cook, *Facing Facts in the Chesapeake Bay*, Environmental Working Group, September 2009, http://www.ewg.org/files/chesapeake-bay-pollution.pdf.

[4] Examples of congressional press releases and letters regarding EPA may be found here: Rep. Frank Lucas -http://www.house.gov/ htbin/blog_inc?BLOG,ok03_ lucas,blog,999,All,Item %20not%20found,ID= 100305_3660,TEMPLATE=postingdetail.shtml; Rep. Collin Peterson -http://collinpeterson.house.gov/press/111th/Peterson% 20calls%20for% 20 bipartisan%20action%20against%20the%20EPA.html; Sen. Inhofe and Sen. Snowe - http://epw.senate.gov/public/index.cfm?FuseAction=Minority.Blogs; and Sen. Chambliss and Sen. Roberts - http://chambliss.senate.gov /public/index.cfm?p=PressReleases&

ContentRecord_id=f7607094-43ca-45c6-a789- 3f91e18e1cca&ContentType_id=5c81ba67-be20-4229-a615-966ecb0ccad6&Group_id=29a81778-8944-46e0-a550- 9d034534e70a.

[5] The Wall Street Journal, "The EPA Permitorium," editorial, November 22, 2010.

[6] American Farm Bureau, "EPA's Regulatory Diet is Unhealthy for America," press release, February 2011, http://fb.org/index.php?fuseaction=newsroom.agenda.

[7] For example, U.S. Congress, Senate Committee on Agriculture, Nutrition, and Forestry, *Oversight Hearing to Examine the Impact of EPA Regulation on Agriculture*, 111th Cong., 2nd sess., September 23, 2010.

[8] For additional information regarding EPA regulations beyond those affecting agriculture, see CRS Report R41561, *EPA Regulations: Too Much, Too Little, or On Track?*.

[9] Some regulations do not become effective immediately. In some cases, the regulation takes effect over time or gradually expands to affect more individuals. Virtually all major EPA regulatory actions are subjected to court challenge, which also delays the implementation.

[10] Court-ordered dates for proposed or promulgated regulations may change. It is not uncommon for EPA to request extensions of time, often due to the need to analyze extensive comments.

[11] 5 U.S.C. §§ 801-808

[12] For additional information on the Congressional Review Act, see CRS Report RL31160, *Disapproval of Regulations by Congress: Procedure Under the Congressional Review Act.*

[13] For example, the House passed H.R. 2454, and the Senate Environment and Public Works Committee reported S. 1733. These bills would have amended the Clean Air Act (CAA) to establish an economy-wide cap-and-trade program for greenhouse gases (GHGs), established a separate cap-and-trade program for HFCs, preserved EPA's authority to regulate GHG emissions from mobile sources while setting deadlines for regulating specific mobile source categories, and required the setting of New Source Performance Standards for uncapped major sources of GHGs.

[14] For example, amending the CAA to revoke some existing regulatory authority as it pertains to GHGs while establishing new authority designed specifically to address their emissions is the approach advocated by the Administration and, indeed, by many participants in the climate debate regardless of their position on EPA's regulatory initiatives. For additional discussion, see CRS Report R41212, *EPA Regulation of Greenhouse Gases: Congressional Responses and Options.*

[15] For additional information on committee jurisdiction, see CRS Report 98-175, *House Committee Jurisdiction and Referral: Rules and Practice* and CRS Report 98-242, *Committee Jurisdiction and Referral in the Senate.*

[16] For example, in the 111th Congress a Manager's Amendment to major climate change legislation added a full title specifically directed toward agriculture. Title V of H.R. 2454, created an offset program for agriculture and forestry related practices to be run by USDA, rather than EPA.

[17] An example of freestanding legislation proposed in the 111th Congress was S. 3072 and its House counterpart, H.R. 4753. These bills, entitled the Stationary Source Regulations Delay Act, provided that during the two-year period beginning on the date of their enactment, EPA could not take any action under the Clean Air Act with respect to any stationary source permitting requirement or any requirement under the New Source Performance Standards section of the act relating to carbon dioxide or methane.

[18] P.L. 111-88, for additional information, see CRS Report R41149, *Environmental Protection Agency (EPA): Appropriations for FY2011.*

[19] P.L. 111-322.

[20] For additional information, see CRS Report RL34069, *Technical Assistance for Agriculture Conservation*.

[21] P.L. 110-161.

[22] U.S. Environmental Protection Agency, "Mandatory Reporting of Greenhouse Gases," 74 *Federal Register* 56260, October 30, 2009.

[23] Department of the Interior, Environment, and Related Agencies Appropriations Act, 2010, P.L. 111-88.

[24] U.S. Environmental Protection Agency, "Mandatory Reporting of Greenhouse Gases," 74 *Federal Register* 56339, October 30, 2009.

[25] U.S. Environmental Protection Agency, U.S. Department of Transportation, "Light-Duty Vehicle Greenhouse Gas Emission Standards and Corporate Average Fuel Economy Standards; Final Rule," 75 *Federal Register* 25324-25728, May 7, 2010.

[26] U.S. Environmental Protection Agency, "Prevention of Significant Deterioration and Title V Greenhouse Gas Tailoring Rule; Final Rule," 75 *Federal Register* 31514, June 3, 2010.

[27] EPA Briefing on the Tailoring Rule, House Energy and Commerce Committee, May 14, 2010. This issue is also discussed in RTI International, for U.S. EPA, "Regulatory Impact Analysis for the Final Prevention of Significant Deterioration and Title V Greenhouse Gas Tailoring Rule," Final Report, May 2010, pp. 64-66, at http://www.epa.gov/ttn/ecas/regdata/RIAs/riatailoring.pdf. A key reason that agricultural sources would not require permits is that EPA excludes what are called "fugitive emissions" from the emissions used to determine whether an agricultural source is a major source subject to permit requirements. Fugitive emissions are emissions that are not released through a stack or vent, or could not be reasonably collected and released through a stack or vent.

[28] P.L. 111-88.

[29] U.S. Environmental Protection Agency, "National Emission Standards for Hazardous Air Pollutants: Reciprocating Internal Combustion Engines; Final Rule," 75 *Federal Register* 9648, March 3, 2010.

[30] U.S. Environmental Protection Agency, "National Emission Standards for Hazardous Air Pollutants for Reciprocating Internal Combustion Engines; Final Rule," 75 *Federal Register* 51570, August 20, 2010.

[31] U.S. Environmental Protection Agency, "National Ambient Air Quality Standards for Particulate Matter," 71 *Federal Register* 61144-61233, October 17, 2006 and U.S. Environmental Protection Agency, "Revisions to Ambient Air Monitoring Regulations," 71 *Federal Register* 61236-61238, October 17, 2006. EPA indicated that it would be expanding its research and monitoring programs to collect additional evidence on the differences between coarse particles typically found in urban areas and those typically found in rural areas. Some stakeholders have expressed concern about EPA's monitoring efforts in rural areas and the future implication monitoring results could have on those areas. Currently, EPA has stated that its monitoring efforts to measure PM are primarily research driven for the purpose of establishing necessary scientific criteria, and not for enforcement purposes.

[32] For additional information, see CRS Report RL34762, *The National Ambient Air Quality Standards (NAAQS) for Particulate Matter (PM): EPA's 2006 Revisions and Associated Issues*.

[33] CRS direct communication with the EPA Office of Air Quality Planning and Standards (OAQPS), January 7, 2011.

[34] EPA had previously targeted proposing changes to the standards by February 2011 and October 2011 for final standards, U.S. Environmental Protection Agency, "Federal

Implementation Plans to Reduce Interstate Transport of Fine Particulate Matter and Ozone: Proposed Rule," 75 *Federal Register* 45220, August 2, 2010.

[35] There has been some concern regarding designations in rural areas for the 2006 $PM_{2.5}$ NAAQS. The designated nonattainment areas for the $PM_{2.5}$ are primarily concentrated in the central, mid-Atlantic, and southeastern states east of the Mississippi River, as well as in California, mostly in and around highly populated metropolitan areas. Several states and industry, agriculture, business, and public advocacy groups petitioned the court challenging certain aspects of EPA's revisions. A February 24, 2009, decision by the U.S. Court of Appeals for the D.C. Circuit granted the petitions in part, denying other challenges (*American Farm Bureau Federation v. U.S. EPA*, No. 06-1410, D.C. Cir., February 24, 2009).

[36] See EPA's PM_{10} designations at http://www.epa.gov/air/oaqps/greenbk/pindex.html. Primarily counties in California and the other western states have not yet come into attainment.

[37] See *Policy Assessment for the Review of the PM NAAQS - First External Review Draft* (EPA 452/P-10-003, March 2010) and *Policy Assessment for the Review of the PM NAAQS - Second External Review Draft* (EPA-452/P-10-007, June 2010); U.S. EPA, *Integrated Science Assessment for Particulate Matter (Final Report)* (EPA/600/R-08/139F, December 2009), and *Risk Assessment to Support the Review of the PM Primary National Ambient Air Quality Standards—External Review Draft* (EPA 450/P_09-006, September 2009). Both assessments are available at http://www.epa.gov/ttn/naaqs/standards/pm/s_pm_index.html. U.S. Environmental Protection Agency, "Release of Draft Documents Related to the Review of the National Ambient Air Quality Standards for Particulate Matter," 74 *Federal Register* 46589, September 10, 2009.

[38] EPA Clean Air Advisory Committee (CASAC) Particulate Matter (PM) Review Committee: CASAC Review of Policy Assessment for the Review of the PM NAAQS – Second External Review Draft (June 2010), August 10, 2010, Draft; CASAC Review of Policy Assessment for the Review of the PM NAAQS - First External Review Draft (March 2010), May 17, 2010. See also CASAC's Review of Quantitative Health Risk Assessment for Particulate Matter— Second External Review Draft, (February 2010), April 15, 2010. These documents are available at http://www.epa.gov/ ttn/naaqs/standards/pm/sjm_index.html.

[39] Letters to EPA Administrator Lisa Jackson: joint letter from 21 Senators, July 23, 2010, http://grassley.senate.gov/ about/upload/Agriculture-07-23-10-dust-letter-to-EPA-signed-version-doc.pdf; joint letter from Senators Kent Conrad and Byron Dorgan, and Representative Earl Pomeroy, August 5, 2010, http://conrad.senate.gov/ pressroom /record. cfm? id=327070&.

[40] For additional background on NAAQS, see the previously discussed "National Ambient Air Quality Standards (NAAQS) - Particulate Matter" section.

[41] The damage that crops and vegetation suffer from ozone exposure is cumulative over the growing season. In order to better measure and provide protection against these impacts, EPA staff recommended a new seasonal (3-month) average for the secondary standard that would cumulate hourly ozone exposures for the daily 12-hour daylight window. Previously, the secondary standard simply measured the highest individual readings for any 8-hour period. CASAC agreed with this recommendation.

[42] For additional information on the proposed standards, see CRS Report R41062, *Ozone Air Quality Standards: EPA's Proposed Revisions.*

[43] For additional information, see CRS Report RL33691, *Animal Waste and Hazardous Substances: Current Laws and Legislative Issues.*

[44] U.S. Environmental Protection Agency, "CERCLA/EPCRA Administrative Reporting Exemption for Air Releases,"73 *Federal Register* 76948-76960, December 18, 2008.

[45] Marc Ribaudo and Robert Johansson, *Agricultural Resources and Environmental Indicators, 2006 Edition*, USDA, ERS, Economic Information Bulletin No. (EIB-16), Washington, DC, July 2006, p. 2.2, http://www.ers.usda.gov/ publications/AREI/EIB16/Chapter2/2.2/. Periodically EPA conducts a National Water Quality Inventory that provides a general water quality assessment based on state collected data. The information for the EPA Inventory is for a relatively small subset of the nation's total waters that are assessed by states and does not represent the waterbodies that were not assessed. For additional information, see EPA, *National Water Quality Inventory: Report to Congress, 2004 Reporting Cycle*, EPA 841-R-08-001, Washington, DC, January 2009, http://water. epa.gov/ lawsregs/guidance/cwa/305b/2004report_index.cfm.

[46] Nonpoint source pollution generally refers to polluted runoff from farms, ranches, forests and urban areas. Nonpoint sources are also subject to natural variability (e.g., weather related events) and depend on many site-specific conditions, such as topography, soil type, and climate.

[47] Much of the federal response to water quality concerns for agriculture is primarily voluntary and incentive-based.

[48] U.S. Environmental Protection Agency, "Revised National Pollutant Discharge Elimination System Permit Regulation and Effluent Limitations Guidelines for Concentrated Animal Feeding Operations in Response to the Waterkeeper Decision, Final Rule," 73 *Federal Register* 225, November 20, 2008, pp. 70417-70486. For additional information on EPA's response to the court decision, see CRS Report RL33656, *Animal Waste and Water Quality: EPA's Response to the Waterkeeper Alliance Court Decision on Regulation of CAFOs.*

[49] The rule specifies thresholds above which permits are required, such as animal feeding operations that stable or confine more than 700 dairy cows, 2,500 swine weighing 55 pounds or more, or 500 horses.

[50] *National Pork Producers Council v. EPA*, 5th Cir., No. 08-61093.

[51] Executive Order 13508, "Chesapeake Bay Protection and Restoration," 74 *Federal Register* 23099-23104, May 15, 2009.

[52] For information, see http://www.chesapeakebay.net/news_federalstrategy. aspx? menuitem =51207.

[53] For background information, see CRS Report 97-831, *Clean Water Act and Total Maximum Daily Loads (TMDLs) of Pollutants.*

[54] For information on the TMDL, see http://www.epa.gov/chesapeakebaytmdl/.

[55] Notice of the TMDL appeared in the *Federal Register* January 5, 2011. U.S. Environmental Protection Agency, "Clean Water Act Section 303(d): Notice for the Establishment of the Total Maximum Daily Load (TMDL) for the Chesapeake Bay," 76 *Federal Register* 549-550, January 5, 2011.

[56] *American Farm Bureau Federation and Pennsylvania Farm Bureau v. U.S. EPA*, Case No. 11-cv-0067 (United States District Court for the Middle District of Pennsylvania 2011).

[57] U.S. Environmental Protection Agency, "Water Quality Standards for the State of Florida's Lakes and Flowing Waters; Final Rule," 75 *Federal Register* 75762-75807, December 5, 2010.

[58] Section 311(j)(1) of CWA

[59] And in accordance with Executive Order 11735 (August 3, 1973), granting EPA the authority to regulate nontransportation-related onshore and offshore facilities.

[60] U.S. Environmental Protection Agency, "Oil Pollution Prevention: Non-Transportation Related Onshore and Offshore Facilities," *Federal Register*, vol. 38, no. 237 (December 11, 1973), pp. 34164-34170.

[61] Per EPA SPCC regulations, "oil," means oil of any kind or in any form, including, but not limited to: petroleum; fuel oil; sludge; oil refuse; oil mixed with wastes other than dredged spoil; fats, oils or greases of animal, fish, or marine mammal origin; vegetable oils, including oil from seeds, nuts, fruits, or kernels; and other oils and greases, including synthetic oils and mineral oils. 40 C.F.R. § 112.2.

[62] Some of the definitions for the terms used to determine SPCC applicability may be subject to interpretation. For example, the definition of "navigable waters" has been a subject of debate and litigation in recent years. See CRS Report RL33263, *The Wetlands Coverage of the Clean Water Act (CWA) Is Revisited by the Supreme Court: Rapanos v. United States.*

[63] Although the definition of facility does not specifically mention farms, farms are explicitly defined as "a facility on a tract of land devoted to the production of crops or raising of animals, including fish, which produced and sold, or normally would have produced and sold, $1,000 or more of agricultural products during a year." See 40 C.F.R. §112.2.

[64] Only counting containers greater than 55 gallons. 40 C.F.R. § 112.1(d).

[65] P.L. 101-380; 33 U.S.C. § 2701 *et seq.*

[66] U.S. Environmental Protection Agency, "Oil Pollution Prevention and Response; Non-Transportation-Related Onshore and Offshore Facilities: Final Rule," 67 *Federal Register* 47041, July 17, 2002.

[67] These actions were, at least in part, related to legal challenges that followed the 2002 final rule.

[68] U.S. Environmental Protection Agency, "Oil Pollution Prevention; Spill Prevention, Control, and Countermeasure (SPCC) Rule - Compliance Date Amendment," 75 *Federal Register* 63093, October 14, 2010.

[69] As of the date of this report, Members of Congress have introduced at least one proposal addressing this issue (e.g., S. 104, introduced by Senator Johanns January 25, 2011).

[70] EPA maintains that the definition of oil "extends to all types of oils in any form." See 67 U.S. Environmental Protection Agency, "Oil Pollution Prevention and Response; Non-Transportation-Related Onshore and Offshore Facilities: Final Rule," 67 *Federal Register* 47042, July 17, 2002. EPA further stated that "milk typically contains a percentage of animal fat, which is a non-petroleum oil" and is thus subject to SPCC provisions (U.S. Environmental Protection Agency, "Oil Pollution Prevention; Spill Prevention, Control, and Countermeasure Rule Requirements –Amendment: Proposed Rule," 74 *Federal Register* 2461, January 15, 2009).

[71] Ibid. To qualify for the exemption, the milk units must meet specific construction standards.

[72] U.S. Environmental Protection Agency, "Oil Pollution Prevention; Spill Prevention, Control, and Countermeasure Plan Requirements – Amendment: Final Rule," 71 *Federal Register* 77266, December 26, 2006.

[73] On December 5, 2008, EPA amended the SPCC rule to clarify certain provisions (U.S. Environmental Protection Agency, "Oil Pollution Prevention; Spill Prevention, Control, and Countermeasure Rule Requirements – Amendments: Final Rule," 73 *Federal Register* 74236, December 5, 2008.). These requirements were to become effective on February 3, 2009. However, the incoming Obama Administration delayed the effective date of the December 2008 rulemaking for regulatory review.

[74] 40 C.F.R. § 112.7(h).

[75] U.S. Environmental Protection Agency, "Oil Pollution Prevention; Spill Prevention, Control, and Countermeasure (SPCC) Rule - Amendments," 74 *Federal Register* 58784, November 13, 2009.

[76] Ibid, p. 58805.

[77] For more information on agriculture-based biofuels, see CRS Report R41282, *Agriculture-Based Biofuels: Overview and Emerging Issues.*

[78] Examples of agriculture-based biofuels policy proponents include organizations who currently benefit directly from policies, such as the National Corn Growers Association (corn-based ethanol) and American Soybean Association (soybean-based biodiesel). Critics include organizations who rely on current biofuel sources for other non-fuel purposes, such as the National Cattleman's Beef Association and National Pork Producers Council.

[79] P.L. 94-163.

[80] P.L. 110-140.

[81] See CRS Report RS22665, *The Supreme Court's Climate Change Decision: Massachusetts v. EPA.*

[82] U.S. Environmental Protection Agency and National Highway Traffic Safety Administration, "Proposed Rulemaking to Establish Light-Duty Vehicle Greenhouse Gas Emission Standards and Corporate Average Fuel Economy Standards; Proposed Rule," 74 *Federal Register* 49454-49789, September 28, 2009.

[83] U.S. Environmental Protection Agency and National Highway Traffic Safety Administration, "Light-Duty Vehicle Greenhouse Gas Emission Standards and Corporate Average Fuel Economy Standards; Final Rule," 75 *Federal Register* 25324-25728, May 7, 2010.

[84] U.S. Environmental Protection Agency and National Highway Traffic Safety Administration, "2017 and Later Model Year Light Duty Vehicle GHG Emissions and CAFE Standards; Notice of Intent," 75 *Federal Register* 62739-62750, October 13, 2010.

[85] U.S. Environmental Protection Agency and National Highway Traffic Safety Administration, "Greenhouse Gas Emissions Standards and Fuel Efficiency Standards for Medium- and Heavy-Duty Engines and Vehicles; Proposed Rule," 75 *Federal Register* 74152-74456, November 30, 2010.

[86] U.S. Environmental Protection Agency and National Highway Traffic Safety Administration, "Proposed Rulemaking to Establish Light-Duty Vehicle Greenhouse Gas Emission Standards and Corporate Average Fuel Economy Standards; Proposed Rule," 74 *Federal Register* 49468, September 28, 2009.

[87] U.S. Environmental Protection Agency and National Highway Traffic Safety Administration, "2017 and Later Model Year Light Duty Vehicle GHG Emissions and CAFE Standards; Notice of Intent," 75 *Federal Register* 62739-62750, October 13, 2010.

[88] U.S. Environmental Protection Agency, *EPA, DOT and California Align Timeframe for Proposing Standards for Next Generation of Clean Cars*, Washington, DC, January 24, 2011, http://yosemite.epa. gov/opa/admpress.nsf/1e5ab1124055f3b28525781f004 2ed40/ 6f34c8d6f2b11e5885257822006f60c0!OpenDocument.

[89] In MY2008, EPA estimates a purchase price increase of $400 for vocational trucks, $1,400 for heavy-duty pickups and vans, and $6,200 for combination tractors (tractor-trailers).

[90] U.S. Environmental Protection Agency, EPA and NHTSA Propose First-Ever Program to Reduce Greenhouse Gas Emissions and Improve Fuel Efficiency of Medium- and Heavy-Duty Vehicles: Regulatory Announcement, EPA-420- F-10-901, Washington, DC, October 2010, http://www.epa.gov/otaq/climate/regulations/420f10901.htm.

[91] http://yosemite.epa.gov/opei/rulegate.nsf/byRIN/2060-AP61, accessed January 26, 2011.

[92] P.L. 109-58.

[93] For more information, see CRS Report R40168, *Alternative Fuels and Advanced Technology Vehicles: Issues in Congress.*

[94] For more information, see CRS Report R40460, *Calculation of Lifecycle Greenhouse Gas Emissions for the Renewable Fuel Standard (RFS).*

[95] Environmental Protection Agency, *Regualtion of Fuels and Fuel Additives: Changes to Renewable Fuel Standard Program*, EPA-HQ-OAR-2005-0161, Washington, DC, February 3, 2010, http://www.epa. gov/otaq/renewablefuels/ rfs2-preamble.pdf.

[96] World Agricultural Outlook Board, *World Agricultural Supply and Demand Estimates (WASDE)*, January 12, 2011..

[97] Mid-point of the projected season average farm price range of $4.90 to $5.70 per bushel, *WASDE*, January 12, 2011.

[98] In April 2008, Texas Gov. Rick Perry wrote the EPA seeking a waiver from the federal ethanol mandate, noting its contribution to higher food prices and dire impact on the cattle industry. The waiver request was denied.

[99] For more information, see CRS Report RL34738, *Cellulosic Biofuels: Analysis of Policy Issues for Congress*.

[100] Environmental Protection Agency, "Regulation of Fuels and Fuel Additives; 2011 Renewable Fuel Standards," 75 *Federal Register* 76790-76830, December 9, 2010.

[101] The practical limitation is likely lower, perhaps 13 to 14 billion gallons.

[102] For more information see CRS Report R40445, *Intermediate-Level Blends of Ethanol in Gasoline, and the Ethanol "Blend Wall"*.

[103] Environmental Protection Agency, "Partial Grant and Partial Denial of Clean Air Act Waiver Application Submitted by Growth Energy to Increase the Allowable Ethanol Content of Gasoline to 15 Percent; Decision of the Administrator; Notice," 75 *Federal Register* 68094-68150, November 4, 2010.

[104] Environmental Protection Agency, "Partial Grant of Clean Air Act Waiver Application Submitted by Growth Energy to Increase the Allowable Ethanol Content of Gasoline to 15 Percent; Decision of the Administrator," signed January 21, 2011 (awaiting publication in the *Federal Register*).

[105] The Alliance of Automobile Manufacturers (Alliance), the Association of International Automobile Manufacturers, Inc. (AIAM), the National Marine Manufacturers Association (NMMA), and the Outdoor Power Equipment Institute (OPEI). OPEI, Fact Sheet: E-15 Partial Waiver Legal Challenge, December 17, 2010. The case is *Alliance of Automobile Manufacturers et. al v. Environmental Protection Agency.*

[106] 40 C.F.R. § 156.10(g)(7).

[107] U.S. Environmental Protection Agency, 52 *Federal Register* 13305, April 22, 1987.

[108] U.S. Environmental Protection Agency, "Public Availability of Identities of Inert Ingredients in Pesticides," 74 *Federal Register* 68215-68223, December 23, 2009.

[109] EPA classified the rule as "Other Significant" which means that although it is not expected to have "major" economic effects, it will be scrutinized in accord with various executive orders concerning regulatory review (Unified Regulatory Agenda, Fall 2010, Dec. 20, 2010, p. 303, http://www.regulations.gov/#!documentDetail;D=EPA-HQ-OA2010-1069-0001.)

[110] U.S. Environmental Protection Agency, "Public Availability of Identities of Inert Ingredients in Pesticides," 74 *Federal Register* 68217, December 23, 2009.

[111] For more information on pesticide use and water quality, see CRS Report RL32884, *Pesticide Use and Water Quality: Are the Laws Complementary or in Conflict?*.

[112] *National Cotton Council of America v. U.S. Environmental Protection Agency,* 553 F.3d 927 (6th Cir. 2009).

[113] EPA had expected to issue the final permit in December 2010, but this did not occur. According to the Office of Management and Budget's website, the permit was submitted to OMB for review on December 22, 2010.

[114] The CWA authorizes EPA to delegate NPDES permitting authority to qualified states, and EPA has done so for the majority of states. For this permit, EPA will be the permitting authority in Massachusetts, New Mexico, Oklahoma, Alaska, Idaho, and the District of Columbia and for certain tribal lands.

[115] EPA, Pesticide Spray and Dust Drift, December. 2009, http://www.epa.gov/pesticides/factsheets/spraydrift.htm.

[116] Letter from Rep. Keith Ellison, Member of Congress, Rep. Raul Grijalva, and Rep. Donna Christensen, Member of Congress, et al. to Lisa Jackson, U.S. Environmental Protection Agency Administrator, EPA-HQ-OPP-2009-0628- 0015, November 20, 2009, http://www.regulations.gov/#!documentDetail;D=EPA-HQ-OPP-2009-0628-0015.

[117] Bergeson & Campbell, Commentary, "2011 Predictions for EPA's Office of Chemical Safety and Pollution Prevention," Jan. 3, 2011, http://www.lawbc.com/news/docs/2011/01/010311-fedreg.htm.

[118] Atrazine is the common name for 6-chloro-N2-ethyl-N4-isopropyl-1,3,5-triazine-2,4diamine.

[119] Jack E. Barbash, Gail P. Thelin, Dana W. Kolpin, and Robert J. Gilliom, Distribution of Major Herbicides in Ground Water of the United States, U.S. Geological Survey, Water-Resources Investigations Report 98-4245 Sacramento, California, 1999, http://water.usgs.gov/nawqa/pnsp/pubs/wrir984245/wrir984245.pdf.

[120] U.S. Geological Survey, "Commonly Used Herbicide Adversely Affects Fish Reproduction," USGS Newsroom, http://www.usgs.gov/newsroom/article.asp?ID=2467&from=news_side.

[121] Richard P. Keigwin, Jr., Director, EPA Pesticide Re-evaluation Division, EPA response to Szmuszkovicz letter regarding the special review status of EPA's current Atrazine review, Aug. 13, 2010, http://www.regulations.gov/ #!documentDetail;D=EPA-HQ-OPP-2003-0367-0191. Hereafter cited as Keigwin.

[122] EPA, Decision Documents for Atrazine, April 6, 2006, http://www.epa.gov/oppsrrd1/REDs/atrazine_combined_docs.pdf.

[123] U.S. Environmental Protection Agency, "EPA Begins New Scientific Evaluation of Atrazine," press release, October 7, 2009, http://yosemite.epa.gov/opa/admpress.nsf /d0cf6618525a9efb85257359003fb69d/ 554b6abea 9d0672f85257648004a88c1! Open Document.

[124] EPA, U.S. Environmental Protection Agency Presentation of the Approach to Reevaluate Atrazine, Nov. 3, 2009, http://www.epa.gov/scipoly/sap/meetings /2009/november/ 110309 meetingtranscripts.pdf.

[125] For more information on the Agricultural Health Study, see http://www. aghealth. nci.nih.gov/.

[126] Keigwin, ibid.

[127] Act of December 28, 1973, P.L. 93-205, 87 Stat. 884. 16 U.S.C. §§ 1531-1544. For a more detailed discussion of ESA and its structure, see CRS Report RL31654, *The Endangered Species Act: A Primer.*

[128] Fish and Wildlife Service and National Marine Fisheries Service, "Interagency Cooperation Under the Endangered Species Act," 74 *Federal Register* 20421-20423, May 4, 2009. The change amended 50 C.F.R. § 402.

[129] For more on the temporary change and issues surrounding its issuance and withdrawal, see CRS Report RL34641, *Changes to the Consultation Regulations of the Endangered Species Act (ESA).*

[130] 16 U.S.C. § 1532. Harassment and harm are further defined by regulation at 50 C.F.R. § 17.3.

[131] See 16 U.S.C § 1538(a)(2).

[132] 16 U.S.C. § 1536(c).

[133] 50 C.F.R. § 402.12(a). Informal consultations are also important, and may be as simple as a federal official of one agency calling an FWS or NMFS official to describe a small project and to find out whether there are any listed species in the vicinity.

[134] 16 U.S.C. § 1536(a).

In: Environmental Considerations in Agriculture ISBN: 978-1-61470-077-7
Editors: E.A. Sargsyan, A.J. Ochion 2011 Nova Science Publishers, Inc.

Chapter 2

AGRICULTURE AND GREENHOUSE GASES*

Claudia Copeland, Megan Stubbs and Kelsi Bracmort

SUMMARY

This report examines the implications for agriculture of the ongoing but inconclusive debate about global climate change. In that debate, agriculture's role is multifaceted. Agriculture is both a source of several greenhouse gases (GHGs) and a "sink" for absorbing carbon dioxide, the most common GHG, thereby partly offsetting emissions. Overall, agriculture is a comparatively modest source of U.S. GHG emissions: it accounts for approximately 7% of U.S. emissions, while sectors such as transportation and electricity generation account for much larger shares. Agriculture's GHG emissions are principally in the form of methane and nitrous oxides emissions.

Whatever the current or future Congresses may do regarding climate legislation, interest in existing and prospective responses by government and others will continue. Administration efforts to develop policies and strategies to address GHGs and climate change have been underway for some time. Two actions by the Environmental Protection Agency (EPA) have drawn the attention of the agriculture industry. One is regulating emissions of GHGs under the Clean Air Act (CAA) and subsequent GHG emission standards for new motor vehicles which, in turn, trigger certain

* This is an edited, reformatted and augmented version of a Congressional Research Service 7-5700, R 41530 publication, dated December 10, 2010.

CAA permitting requirements. A second, related action is a rule to require reporting of GHG emissions by certain facilities. Regarding both, EPA took steps to focus on the largest emitters and ensure that few agricultural sources would be subject to new GHG requirements. Still, EPA's overall initiatives have been widely criticized, and the 111th Congress intervened through a funding bill to largely exclude agriculture from EPA's regulatory requirements. During the 111th Congress, the House passed a comprehensive climate change bill (H.R. 2454), and a Senate committee reported a companion (S. 1733). Although no legislation was enacted, both bills included provisions excluding agriculture from regulatory requirements and promoting agricultural practices to reduce or offset emissions from regulated sources.

Traditionally, practices such as conservation tillage have been used for soil conservation and water quality improvement, but their value for climate change abatement or mitigation is receiving increased attention. A number of strategies, technologies, and practices exist to reduce methane and nitrous oxides emissions at the farm level, but implementation faces financial and monitoring challenges.

Programs administered by the U.S. Department of Agriculture (USDA) provide financial incentives and technical assistance to encourage implementation of certain farming practices. While the focus of most programs is not on GHG emission reduction, USDA is giving greater attention to GHGs in administering its suite of existing programs.

Results of the 2010 congressional elections have altered political dynamics in Congress on many issues, and leadership of both political parties have indicated that neither currently plans to pursue comprehensive approaches to addressing climate change in the 112th Congress, although some elements of previous proposals may move through the legislative process. How agriculture fits in these discussions—both as a source of GHG emissions and contributions that the sector can make to mitigating climate change—has drawn interest in the past and likely will do so again.

INTRODUCTION

This report examines the implications for agriculture of the ongoing but inconclusive debate about global climate change.[1] Whatever the current or future Congresses may do regarding climate legislation, interest in existing and prospective governmental and private sector responses will surely continue. Existing governmental climate change activities affecting or potentially affecting agriculture include the Environmental Protection Agency's (EPA) initiative under the Clean Air Act to address emissions of greenhouse gases (GHGs); U.S. Department of Agriculture (USDA) conservation programs that can encourage practices affecting agriculture's emissions or sequestration of GHGs; and provisions of law encouraging the use of biofuels.[2] Private sector concerns about climate change also could affect agriculture. For example, some companies increase investment in drought-tolerant seed technology in response to water resource pressures and potentially longer periods of extreme temperatures resulting from a changing climate.

In the climate change debate, agriculture's role is multifaceted. Agriculture is both a source of several GHGs and a "sink" for absorbing carbon dioxide, the most common GHG, thereby partly offsetting emissions. Overall, agriculture is a comparatively modest source of U.S. GHG emissions: it accounts for approximately 7% of U.S. emissions, while transportation accounts for 27% and electricity generation for 35%. Further, it should be recognized that the data on various agriculture emissions are of varying precisions and that the implications of land use changes for emissions are not well understood and are hard to measure. Similarly, agriculture is a dynamic activity and emissions/sequestration can change and also can be manipulated to some degree.

During the 111th Congress, comprehensive climate change legislation passed the House and was reported by a Senate committee, but no comprehensive bill was enacted. At the same time, the Administration moved forward on several fronts to address climate change, including research, EPA's clean air initiative, and USDA's promotion of conservation practices that can diminish emissions. As these various initiatives progressed, the agriculture community weighed in at several points. One outcome was that the comprehensive climate change bills largely excluded agriculture from regulatory requirements, and another was that Congress through a funding bill excluded agriculture from certain EPA regulatory requirements concerning reporting of emissions (e.g., from manure management practices).

This report places in context federal agricultural policymaking with respect to the climate change debate. It describes agricultural activities involving cropland and livestock production that contribute to emissions and sequestration of GHGs; it reviews recent climate change initiatives, agricultural stakeholder responses, and current status; and it summarizes the implications of ongoing federal programs affecting agricultural emissions and sequestration.

AGRICULTURAL EMISSIONS AND SEQUESTRATION OF GREENHOUSE GASES

Agricultural activities can be both a source and a "sink" for GHGs, releasing several GHGs through plant and animal respiration and plant decomposition and removing carbon dioxide (CO_2) through photosynthesis, thus storing/sequestering it in vegetation and soils.[3] Animal agriculture contributes directly to emissions of GHGs through a variety of processes such as enteric fermentation in domestic livestock (i.e., digestion) and manure management systems and practices. Non-livestock source categories in agriculture likewise emit GHGs, including rice cultivation, agricultural soil management, and field burning of agricultural residues. A range of land management, agricultural conservation, and other farmland practices also can reduce or abate emissions and/or sequester carbon to some extent. These include soil conservation, manure and grazing management, and land retirement, conversions, and restoration.

As shown in *Table 1,* agriculture is estimated to have emitted 6.9% of total U.S. greenhouse gas emissions in 2008, including CO_2, methane (CH_4), and nitrous oxides (N_2O). Conversely, agriculture is estimated to have been a sink for 5.4% of total GHG emissions that were sequestered in the United States. Unlike other prominent economic sectors (e.g., electricity generation and transportation), agriculture sector emissions are dominated by CH_4 and N_2O, not CO_2.[4]

Carbon Dioxide

Billions of tons of carbon in the form of CO_2 are emitted to the atmosphere annually from anthropogenic sources and natural processes. Some agricultural land management uses and practices involve both emissions and removal of CO_2 from the atmosphere, a combination that is referred to as CO_2 flux. Agriculture directly emits CO_2 from fossil fuel combustion by transportation and other on-farm activities; fossil fuel combustion by agricultural sources accounted for an estimated 45.4 million metric tons of CO_2 equivalent ($MMTCO_2e$) in 2008, or 76% of the CO_2 emissions shown in Table 1.

Table 1. Estimated GHG Emissions and Sequestration: U.S. Agriculture and U.S. Total, All Sources (2008)

Million Metric Tons of CO_2 equivalent ($MMTCO_2e$)

Source	Emissionsa	Sequestrationb	Net
Agriculture CO_2	*59.9*	*(51.0)*	*8.9*
Agriculture CH_4	*194.0*	—	*194.0*
Agriculture N_2O	*233.2*	—	*233.2*
Agriculture Subtotal	480.5	(51.0)	429.5
U.S. Total, All Sources	6,956.8	(940.3)	6,016.4
% U.S. Total, Agriculture	6.9%	5.4%	—

Source: U.S. Environmental Protection Agency, Inventory of U.S. Greenhouse Gas Emissions and Sinks: 1990-2008, April 15, 2010, EPA 430-R-10-006, Tables ES-7, 7-1, 2-12; calculations by CRS.

a) Includes CO_2, CH_4, and N_2O emissions attributable to the agriculture sector, including land use/land use changes and fossil fuel combustion from transportation and other on-farm activities. Excludes emissions from forestry activities and emissions allocated to electricity generation for agricultural activities.

b) Measured agricultural sequestration categories include cropland remaining as cropland, grassland remaining as grassland, land converted to cropland, and land converted to grassland. Excludes sequestration by forestry activities.

Agriculture soils also emit relatively small amounts of carbon through the application of liming and urea fertilizer (7.6 $MMTCO_2e$ in 2008). The U.S. agricultural sector is a minor source of CO_2 emissions—1% in 2008—while electricity generation accounts for 40%, and transportation accounts for 30%. On the other hand, agricultural soils sequester nearly nine times more carbon than they emit. Soil carbon sequestration is largely due to conversion of cropland, an increase in adoption of conservation tillage practices that preserve soil carbon, and an increase in the amount of organic fertilizer (manure and sewage sludge) that farmers apply to croplands in place of synthetic fertilizers.[5]

EPA's Greenhouse Gas Inventory reports on four categories of agricultural land use practices. Three of these categories—cropland remaining as cropland, grassland remaining as grassland, and land converted to grassland—demonstrated carbon sequestration in 2008, while the fourth—land converted to cropland—contributed emissions of about 6 $MMTCO_2e$, not sequestration. Carbon sequestration on farm lands is currently estimated to contribute more than 5% of total sequestration by all sources (which occurs principally through forested lands remaining forest land) and to mitigate less than 1% of total annual GHG emissions in the United States (forested lands mitigate about 11% of total annual GHG emissions in the United States).

There is ongoing debate about the permanence, or duration, of many carbon sequestration practices. Permanence depends on the practice itself and such site-specific factors as location, climate and land condition. For example, reforestation and similar forestry activities may be capable of long term emission reduction (from 20 to 200 years). However, cropping practices such as conservation tillage or other cropland changes (e.g., transitioning to improved pasture) may or may not be long term, depending on how long a farmer maintains the practice, whether a farmer receives and continues to receive financial or technical assistance as incentive for maintaining the practice, whether the farmer is able to withstand lower yields in the near-term, or whether high prices shift idle land back into production.[6]

Methane and Nitrous Oxides

The two principal GHGs emitted by agricultural sources are CH_4 and N_2O. An estimated one-half of global CH_4 comes from anthropogenic sources (i.e., from human activities), of which agriculture is the largest source; livestock production is a major component within the sector.[7] EPA estimates that nearly one-third of U.S. CH_4 emissions come from livestock. Agricultural CH_4 is produced by ruminant animals, but it also is emitted during microbial degradation of organic matter under anaerobic conditions. Nitrous oxides are formed as a result of crop fertilization practices, directly via the microbial processes of nitrification and denitrification, and indirectly via volatilization and surface water runoff and leaching into ground water.

In 2008, total anthropogenic emissions of CH_4 and N_2O from all U.S. sectors were 885.8 $MMTCO_2e$, while total agricultural sector GHG emissions of these two GHGs totaled 427.5 $MMTCO_2e$, or 48% of total U.S. emissions of both GHGs.[8] As shown in *Figure 1*, agricultural emissions of CH_4 are predominantly from enteric fermentation in domestic livestock, with lesser contributions by livestock manure management, rice cultivation, and field burning of agricultural residues. Over 90% of agricultural emissions of N_2O are from agricultural soil management, with small contributions by livestock manure management and field burning of agricultural residues.[9]

Million Metric Tons of CO_2 equivalent (MMTCO_2e)

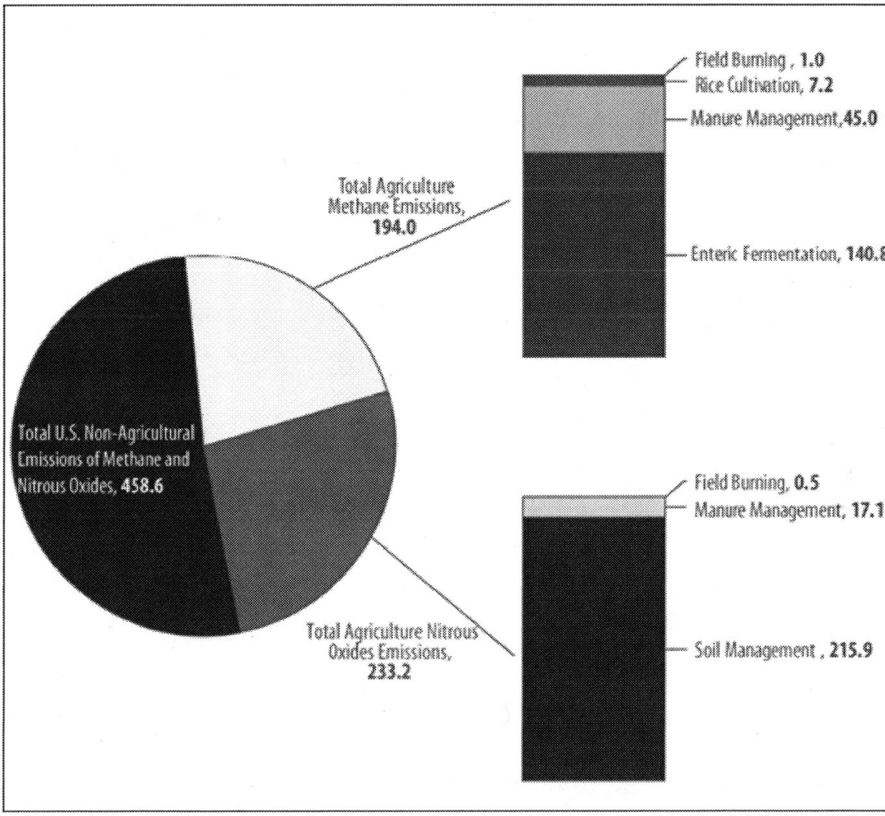

Source: Prepared by CRS from U.S. Environmental Protection Agency, Inventory of U.S. Greenhouse Gas Emissions and Sinks: 1990-2008, April 15, 2010,EPA 430-R-10-006, Tables 6-1, 2-1.

Figure 1. Estimated Emissions of Methane and Nitrous Oxides: U.S. Total and Agricultural Sector, 2008.

Methane Emissions from Agriculture

Livestock sources were responsible for one-third of all U.S. CH_4 emissions in 2008. More than three-quarters of CH_4 emissions from livestock resulted from normal digestive processes (i.e., enteric fermentation[10]) in ruminant animals such as cattle, sheep, and goats. Cattle account for the majority of CH_4 emissions from U.S. livestock (because of their large population, large animal size, and particular digestive characteristics), and emissions changes over time tend to track changes in beef and dairy cattle populations. Feed quality and the amount of feed intake by animals also affect CH_4 emissions.[11]

The management of livestock manure also can produce CH_4. Methane is produced from the decomposition of liquid-based livestock manure that is stored or treated in lagoons, ponds, tanks, or pits. Factors that affect the amount of CH_4 produced include temperature, moisture, and storage time, because they influence the growth of bacteria that are responsible for CH_4 formation. An animal's feed also can be a factor: in general, the greater the energy content of the feed, the greater the potential for methane emissions.

When manure is handled as a solid or deposited on pasture, range, or paddock lands, it produces little or no CH_4, and in fact, the majority of manure in the United States is handled as a solid. However, the general trend in manure management, particularly for dairy and swine, is towards use of liquid systems. According to EPA, states such as California, New Mexico, and Idaho have seen increases in dairy populations as the industry becomes more concentrated, along with greater use of liquid-based systems to manage and store manure. A consequence of the dairy industry's shift toward larger facilities translates into an increasing use of liquid manure management systems, which have higher potential CH_4 emissions than dry systems. Between 1990 and 2008, methane emissions from manure management increased on average 2.5% annually. The majority of this increase was from swine and dairy cow manure, where emissions increased 50% and 91%, respectively.[12]

Rice cultivation and field burning of agricultural residues also contribute CH_4 emissions, but they are small contributors of U.S. emissions (1.5% of all CH_4 emissions). Crop residue burning also produces small amounts of both CH_4 and N_2O (see *Figure 1* and *Table A-1*). Although field burning is not a widely used practice of farmers for disposing of crop residues, it is used throughout the United States for disposal of about 3% of the residue of wheat, rice, sugarcane, corn, barley, soybeans, and peanuts. According to EPA, annual emissions from this source have remained relatively constant since 1990.[13]

Nitrous Oxides Emissions from Agriculture

Nitrous oxides emissions are produced by biological processes that occur in soil and water and by a variety of human activities involving agriculture, energy, industry, and waste management. Agricultural soil management practices (e.g., fertilizer application and other cropping practices) produce the majority of N_2O emissions in the United States and accounted for nearly 68%

of U.S. N_2O emissions in 2008.[14] Year to year fluctuations in agricultural soil emissions of N_2O reflect variations in weather patterns, synthetic fertilizer use, and crop type.[15]

A number of agricultural activities increase mineral nitrogen availability in soils, thereby increasing the amount available for the microbial processes that produce nitrous oxide. Direct increases in soil mineral nitrogen occur as a result of practices such as fertilization, application of manure on soils, and production of nitrogen-fixing crops and forages such as clover and alfalfa. Direct N_2O emissions from croplands tend to be high in the Corn Belt states where highly fertilized corn and nitrogen-fixing soybean crops are grown. Direct emissions also tend to be high from grasslands in the central and western states where a high proportion of land is used for cattle grazing (e.g., emissions by grazing animals whose manure is not managed and from retention of crop residues).

Indirect emissions, which occur when mineral nitrogen is transported from the soil either in gaseous or aqueous forms and later is converted into N_2O, comprise about 25% of emissions from agricultural soil management activities. These types of emissions occur in many of the same U.S. regions as direct emissions (e.g, central and western United States). There are two pathways leading to indirect emissions. The first results from volatilization of nitrogen to the atmosphere (e.g., from nitrogen fertilizer). Indirect emissions also occur through surface transport and runoff from farmland into nearby streams and lakes.

Livestock manure management activities produce N_2O emissions, as well as CH_4. Direct N_2O emissions are released from dry manure handling systems, pasture, solid storage, and—similar to agricultural soil management activities—indirect emissions result from volatilization of nitrogen or runoff of nitrogen during manure treatment, storage, and transportation. Both direct and indirect emissions of N_2O have remained fairly steady since 1990, according to EPA.[16]

Data Uncertainty and Varying Estimates

Scientists have considerable confidence in characterizing U.S. GHG emissions, particularly for major industrial sectors where statistics such as fossil fuel consumption are relatively accurate. At the same time, EPA and others recognize that there are uncertainties associated with some of the emission estimates data, especially for sectors that are smaller contributors of emissions, due to a lack of data or an incomplete understanding of how emissions are generated or may be measured.

Uncertainty is apparent in much of the data on agriculture's emissions that are presented in EPA's Greenhouse Gas Inventory. Emissions estimated in the Inventory may vary from year to year based on new data and changes in assumptions and methodology. Regarding CH_4 emissions from enteric fermentation, which are the main agricultural source of those emissions, the Inventory states that the lower- and upper-bound uncertainties are -11% and +18%. However, regarding other emissions, such as N_2O emissions from soil management, which is the main agricultural source of those emissions, there is greater uncertainty. For example, direct emissions, which account for the majority of N_2O emissions from agricultural soil management, have lower- and upper-bound uncertainties of -24% and +63%, while indirect emissions (i.e., from volatilization) have greater uncertainties due to lack of data, as well as uncertainties regarding major crops and application of manure and other organic fertilizer amendments: the uncertainty range in the report is -48% and +142%.[17]

Notwithstanding specific uncertainties, whether agriculture's emissions are 7% of the U.S. total, as reported in EPA's Inventory, or 3%, or 11%, is not especially critical to policy debates about GHG emissions and climate change. What is evident in the data is that agriculture is a much smaller source of emissions than other economic sectors—especially electricity generation and transportation. Thus, while policy debate about climate change continues to occur internationally, nationally, and regionally, one question is where does agriculture fit in those discussions. Broadly speaking, there are three distinct policy tracks that could define agriculture's role. One option would be to regulate agriculture and other sources of GHGs in order to mitigate or abate emissions. A second would be to promote practices by agricultural sources that may diminish or mitigate the sector's emissions voluntarily. A third option would be to do nothing.

EPA Activities

Efforts have been underway in the Administration to develop policies and strategies to address GHGs and climate change. The 111th Congress, too, considered legislation in this area: comprehensive climate and energy legislation passed the House in July 2009 and was reported by a Senate committee, but no comprehensive bill was enacted. Agriculture generally has been a major part of these discussions, but so far the agriculture sector has been largely excluded from regulatory and legislative proposals. Two sets of actions by the EPA concerning GHG emissions have drawn stakeholders' attention.

Regulating GHGs under the Clean Air Act and the Tailoring Rule

First, in July 2008, the Bush Administration published an Advance Notice of Proposed Rulemaking (ANPR) in connection with its consideration of how it should comply with *Massachusetts v. EPA*, in which the Supreme Court held that the Clean Air Act (CAA) authorizes EPA to regulate emissions from new motor vehicles on the basis of their climate change impacts. The Court held that the EPA Administrator must determine whether or not emissions of GHGs from new motor vehicles cause or contribute to air pollution which may reasonably be anticipated to endanger public health or welfare (i.e., an endangerment finding), or whether the science is too uncertain to make a reasoned decision.[18] Responding to this ruling with the ANPR, EPA discussed a wide range of CAA authorities and programs that could potentially be used to address climate change, including the permitting provisions in Title V of the act.[19] The ANPR did not propose or recommend the use of any particular CAA authority, or commit to specific next steps to address GHGs from any category of emission sources.

Agricultural sources were not specifically referenced in any of this ANPR discussion; nevertheless, agriculture stakeholders—especially many representing livestock operations—were highly critical of potential economic impacts on their operations and the possibility that Title V permits might be required. In the months following the ANPR, EPA officials said that the agency had no plans to tax livestock or pursue other "doomsday scenarios" for new regulations.[20] The public comment period on the ANPR ended in November 2008; no further action on it occurred.

However, in December 2009, EPA Administrator Jackson signed two endangerment findings about GHGs. First, the Administrator found that the current and projected concentrations of six GHGs in the atmosphere (including CH_4 and N_2O) threaten the public health and welfare of current and future generations. Second, the Administrator found that GHG emissions from motor vehicles contribute to the atmospheric concentrations of the six key GHGs and hence to the threat of climate change.[21] The endangerment finding does not itself impose any CAA requirements on industry or other entities or trigger regulation under the entire act. However, the endangerment finding is a prerequisite to finalizing proposed greenhouse gas emission standards for light-duty vehicles, which EPA proposed jointly with the Department of Transportation in September 2009. When EPA's proposed light-duty vehicle rule takes effect (expected to be January 2011), other CAA requirements will be triggered. In particular, stationary sources that emit any of the six GHGs covered by the endangerment finding will be subject to certain permitting requirements under the Title V operating permit and New Source Review (NSR) provisions in the law.[22]

Related to the CAA requirements that are triggered by the endangerment finding and light-duty vehicle rule, on May 13, 2010, EPA issued a rule specifying thresholds for GHG emissions that define when Title V and NSR permits would be required. In the absence of the rule, called the GHG Tailoring Rule, sources that emit as little as 100 tons per year of CO_2 equivalent of GHGs would be subject to CAA permits. In order to limit the number of facilities that would be required to obtain permits, in the Tailoring Rule EPA established a threshold of 100,000 tons per year of CO_2 equivalent of GHG emissions. EPA estimates that the rule will cover 67% of the nation's largest stationary source GHG emitters, while shielding small businesses and agriculture operations from new permitting requirements.[23] EPA believes that livestock and production agriculture operations will not be subject to CAA permitting as a result of the Tailoring Rule, because of the high threshold in the rule and because the rule does not apply to so-called "fugitive emissions" from sources of enteric fermentation and animal manure management systems.[24]

The Tailoring Rule does apply to GHG emissions from internal combustion diesel engine generators, including those used on farms. However, because of the 100,000 tpy threshold in the rule, EPA estimates that no farm stationary fuel combustion sources emit GHGs (i.e., CO_2) at levels that would be subject to the rule.

Mandatory Reporting of Greenhouse Gases

A second EPA action that drew agriculture's attention was an April 2009 EPA proposal to require reporting by certain facilities that emit GHGs and by suppliers of fossil fuels and industrial GHGs. The proposal responded to a congressional directive in the FY2008 Consolidated Appropriations Act (P.L.

110-161) for EPA to develop a comprehensive national system for reporting emissions of CO_2 and other GHGs produced by major U.S. sources. Included in the categories of sources that would be subject to the proposed rule were manure management systems that emit, in the aggregate, methane and nitrous oxide in amounts equivalent to 25,000 metric tons of CO_2 equivalent or more per year. Because of the proposed reporting threshold, EPA initially estimated that fewer than 50 beef cattle, dairy cattle, and swine operations would be subject to the rule; an unknown number of poultry operations also would be covered.[25]

A number of agriculture stakeholders criticized the proposal. Many noted that agriculture as a whole is responsible for only a small percentage of total GHGs and questioned why manure management systems in particular were included in the proposal, since they are responsible for less than 1% of total U.S. GHGs (see *Table A-1*). Other categories of agricultural sources, such as livestock enteric fermentation and soil management, emit larger amounts of CH_4 and N_2O, but were not included in the proposal. EPA explained that the proposal did not include reporting by the other agriculture categories such as field burning of agricultural residues because, for those sources, there are no direct GHG emission measurement methods available except for expensive and complex equipment. Using emissions estimates for such sources, instead of direct measurement, would have a high degree of uncertainty and likely would burden a large number of small emitters, EPA said. Some who commented on the proposal said that similar concerns— about a lack of accurate measurement methods and the costly burden of compliance with only very small benefits—apply equally to reporting by manure management systems.

The EPA Administrator signed the final reporting rule on September 22, 2009.[26] As in the proposal, the final rule applies to manure management facilities with the same reporting threshold of 25,000 metric tpy of CO_2 equivalent of GHGs, but not to other agricultural sources or agricultural land uses. In response to comments about the burden of the rule, EPA removed manure sampling requirements and instead will allow facilities to use default values for estimating emissions. EPA also made certain recalculations of affected facilities and now estimates that approximately 107 livestock facilities will be subject to the reporting rule. The final rule identifies population threshold levels below which facilities are not required to report emissions, such as fewer than 29,300 beef cattle and fewer than 3,200 dairy cattle. These thresholds would exclude 99% of beef feedlots, dairy farms, and others operations with manure management systems.[27] Facilities subject to the rule would report annually beginning in January 2011. However, as discussed below, in passing EPA's FY2010 appropriations legislation (P.L. 111-88), Congress included bill language barring EPA from using funds under that act to implement mandatory GHG reporting by manure management facilities.

CONGRESSIONAL INTEREST

The 111th Congress showed interest in several aspects of issues concerning agriculture and GHGs, acting mainly to exempt or relieve agriculture from potential regulation of sources' GHG emissions.

First, legislation was introduced in the 111th Congress in response to concerns raised by EPA's July 2008 ANPR that the agency might require CAA permits for GHG emissions from agriculture, which some groups characterized as a "cow tax proposal." The legislation, S. 527 and H.R. 1426, would have amended the Clean Air Act to mandate that no Title V permit be issued for controlling carbon dioxide, nitrogen oxide, water vapor, or methane emissions "resulting from biological processes associated with livestock production." In addition, in the FY2010 appropriations bill for EPA (P.L. 111-88), Congress included a provision similar to the prohibitory language of S. 527 and H.R. 1426. As adopted, the measure prohibits EPA from using funds under the act to promulgate or implement any rule requiring the issuance of CAA Title V permits for GHG emissions associated with livestock production.

Second, also in final action on P.L. 111-88, Congress blocked EPA from using funds in the bill to implement any rule that would require mandatory reporting of GHG emissions from manure management operations. This bill language applies to manure management systems of all sizes, not just to those that emit more than 25,000 metric tons of CO_2 equivalent per year, as contained in EPA's mandatory reporting rule. As noted previously, EPA's rule excludes reporting by 99% of farms with manure management systems; P.L. 111-88 excluded the other 1% of operations.

Third, the 111th Congress debated comprehensive climate change bills and in that context considered whether or how to include agriculture and other sources of GHGs in the legislation. In July 2009, the House passed the American Clean Energy and Security Act of 2009 (H.R. 2454), legislation that covers clean energy, energy efficiency, reducing global warming pollution, transitioning to a clean energy economy, and agriculture and forestry related offsets. The complex and controversial legislation reflected compromises on various issues, including a number of negotiated changes sought by agriculture interests.[28] A key feature of the House-passed bill was an economy-wide cap-and-trade system designed to reduce GHG emissions from covered entities. As passed, the legislation excluded any agricultural enterprise or any small business enterprise that emits less than 25,000 tons of CO_2 equivalent of GHG emissions per year. Animal agriculture sources were excluded entirely from the definition of "covered entities" in H.R. 2454, because of their *de minimis* impact on the climate; thus, they would not have been subject to the cap or other mandatory provisions of the bill.[29]

A key feature of H.R. 2454, as passed by the House, was the creation of a carbon offset provision for agriculture. The so-called "Peterson Amendment" was added to H.R. 2454 just prior to the floor debate, following negotiations between the Chairmen of the House Energy and Commerce Committee and the House Agriculture Committee. Among other provisions, the Peterson Amendment allowed for certain agricultural and forestry activities to become eligible to participate in a carbon offset program. Offsets (emission reductions from non-covered sources) could be purchased by covered entities and used to meet their compliance obligations. Thus, the agricultural and forestry sectors could earn income for any emission reductions that it undertakes, provided that the reductions are measurable and verifiable. The legislation also would have established the offset program under USDA (rather than EPA), a key difference sought by agriculture stakeholders.[30]

Comprehensive climate change legislation was reported from the Senate Environment and Public Works Committee in February 2010 (S. 1733, the Clean Energy Jobs and American Power Act). Regarding agriculture, this bill was similar to H.R. 2454 in that it used the same emissions threshold (25,000 metric tons of CO_2 equivalent per year) applicable to the cap-and-trade and other mandatory provisions and would exclude animal agriculture from the definition of "covered entities." Like H.R. 2454, S. 1733 would have allowed for agriculture and forestry offsets as part of a cap-and-trade scheme. Also in the Senate, the Clean Energy Partnerships Act of 2009 (S. 2729) was introduced by Senator Stabenow shortly after the Senate Environment and Public Works Committee completed work on S. 1733. This bill (often referred to as the "Stabenow Amendment") would have expanded the agricultural and forestry carbon offset provisions in the comprehensive climate bills (e.g., S. 1733) and also would have allowed for certain other provisions benefitting U.S. farmers and landowners.

AGRICULTURAL CONSERVATION PRACTICES AND GHGS

Some degree of GHG emissions reduction from agricultural activities can be achieved with existing conservation and land management practices, which also conserve and improve the quality of soil, water, air, energy, and plant and animal life.[31] Thus, in addition to or in place of regulating agricultural emissions, another policy option could be to promote various voluntary on-farm practices—ranging from reduced tillage to prescribed grazing—that could mitigate emissions. The effectiveness, complexity, cost, and break-even point of the conservation practice varies based on the farm type, farm size, land management, and agricultural commodity, among other factors. A list of selected conservation practices that can reduce agricultural GHG emissions is displayed in Table 2.

Typically, conservation practices have been used for soil conservation and water quality improvement, not climate change abatement or mitigation. However, recent investigations have shown certain conservation practices can significantly reduce agriculturally based GHG emissions. Ideal conservation practices that also could reduce GHG emissions are associated with the primary agricultural sources of these emissions (described previously)—enteric fermentation, manure management, and agricultural soil management—or more simply, croplands and livestock. Conservation practices that fall under the domains of nutrient management, tillage operations, field management, precision agriculture, manure management, and dietary management will likely have the most influence on decreasing agricultural GHG emissions. A combination of practices may be necessary in some cases to significantly reduce GHG emissions. However, certain conservation practices may decrease a single GHG while simultaneously increasing others.[32]

Depending on the agricultural source of GHG emissions, different conservation practices would likely need to be used to obtain desired reductions. For instance, GHG control strategies for cropland differ from those used for livestock. Cropland GHG emissions generally are distributed over a vast area of land. Land management practices, sometimes referred to as "non-structural practices," work well to reduce GHG emissions from cropland.[33] On the other hand, livestock GHG emissions may be widely dispersed (emissions originate from a herd of cattle grazing) or may emanate from a point-source (emissions originate from the manure handling system). A mixture of feed management, nonstructural practices, and structural practices may be used to reduce livestock GHG emissions.

In addition to reducing GHG emissions directly, some conservation practices provide another climate change benefit: carbon sequestration (storage), which takes carbon out of the atmosphere. Wetland restoration is a prime example of a conservation practice that sequesters carbon by re-establishing a sustainable ecosystem that provides a relatively greater degree of permanence in CO_2 sequestration than some other practices.[34]

Strategies, technologies, and practices to reduce CH_4 and N_2O emissions at the farm level are not mandated by federal policies. As discussed further below, current federal policies are voluntary and offer incentives for reducing GHG emissions. Absent mandates or increased incentives, financial and monitoring challenges have stalled large-scale adoption of certain practices and technologies. For example, CH_4 emitted from a dairy farm via enteric fermentation (i.e., digestion) and manure management can be reduced by feeding dairy cows a high-quality forage and using an anaerobic digestion system to capture methane from the manure. However, the high cost of anaerobic digestion systems is a significant barrier to their widespread use for methane capture.[35] Similarly, N_2O emitted from corn fields due to nutrient and soil management efforts (e.g., fertilization and tillage) can be reduced with efficient application of fertilizer and conservation tillage, but costs of adopting appropriate technologies and practices can be prohibitive. Overall, problems of quantifying, monitoring, and verifying emission reduction or carbon storage may make it impracticable to include many agricultural activities in some GHG reduction programs, such as trading, because they might not meet credible standards.[36]

USDA Activities and Programs

Existing conservation and farmland management programs administered at the federal and state levels often encourage agricultural practices that can reduce GHG emissions and/or sequester carbon (see *Table 2*). Most of these programs are voluntary and were initiated mainly for other production or environmental purposes (such as soil fertility and water quality improvement). Few existing programs specifically address GHG emission concerns in the agriculture and forestry sectors.

Table 2. Conservation Practices and GHGs

	Practice				GHGs Affected
Category	Activity	Description	CO_2	CH_4	N_2O
Cropland	Nutrient management planning	Timing, rate, and type of fertilizer applications	X	X	X
	Fertilizer land application	Incorporating manure fertilizer, synchronizing N supply with plant need through timing and placement (e.g., precision agriculture)		X	X
	Proper nutrient crediting (manure/crops)	Crediting nutrient inputs from prior year amendments or legumes		X	X
	Residue management	Using reduced or no-till practices reduces	X		X
	Pasture and hay planting	Planting high nutrient species and maintaining permanent, vigorous plant growth	X	X	X
Livestock	Manure management	Relocating manure fertilizer from crop land with nutrient excesses to crop land with nutrient deficiencies; anaerobic digestion technology; covered lagoon with flaring	X		X
	Dietary management	Managing animal diets effectively (e.g., feed additives)		X	X
	Conservation crop rotation	Using legume crops in rotation		X	X

Source: Adapted by CRS from Solomon Haile, Clyde Fraisse, and P.K. Ramachandran Nair, et al., Greenhouse Gas Mitigation in Forest and Agricultural Lands: Reducing Emissions, University of Florida IFAS Extension, AE443, December 2008, http://edis.ifas.ufl.edu/pdffiles/AE/AE44300.pdf. The table reflects no relative order of importance of the cited control strategy or activity.

However, USDA and some states have started to focus additional attention on the potential for emissions reduction and carbon storage under certain existing programs.[37] These include conservation, forestry, energy, and research programs within existing farm legislation.

In general, conservation programs administered by USDA and state agencies encourage farmers to implement certain farming practices and often provide financial incentives and technical assistance to support their adoption. Participation in these programs is voluntary, and long-term maintenance of implemented practices is not mandatory following the completion of a contract. The effectiveness of these practices depends on the type of practice, how well it is maintained, and also on the length of time a practice is undertaken. These programs are generally designed to address site-specific improvements based on a conservation plan developed with the assistance of USDA technical field staff, state extension services, or private technical service providers. Conservation plans consider the goals and land resource base for an individual farmer or landowner and are typically a necessary precursor to participating in USDA's conservation programs.[38] This section describes relevant USDA programs that can provide financial and/or technical assistance for the types of on-farm practices described above to mitigate agriculture's GHG emissions.

Conservation Programs

Most conservation programs administered by USDA are designed to take land out of production and improve it (i.e., land retirement/easement programs) or to improve management practices on land in production (working lands programs).[39] Programs include some level of technical assistance to assist implementation and typically offer a cost-share contract to producers to implement practices necessary to achieve conservation goals. Many of these programs are provided for in Title II of the 2008 farm bill (P.L. 110-246, the Food, Conservation, and Energy Act of 2008).[40]

USDA has expanded some of its existing farmland conservation programs to further encourage agricultural emission reductions and carbon sequestration.[41] As described previously, a number of conservation and land management practices can reduce net emissions directly, and many of the practices are encouraged under working lands programs, such as the Environmental Quality Incentives Program (EQIP) and the Conservation Stewardship Program (CSP). USDA has provided additional technical guidance to make GHG a priority resource concern in working lands programs by giving greater weight to projects that promote anaerobic digestion, nutrient management plans, and other types of cropland practices, such as installing shelter belts and windbreaks, encouraging conservation tillage, and providing resources for biomass energy projects. Several working lands programs list a reduction in emissions as a national priority for the program, which affects the overall funding and ranking of projects. USDA has modified how it scores and ranks offers to enroll land in the Conservation Reserve Program (CRP) in order to place greater weight on installing vegetative covers that sequester carbon. USDA also has an initiative under CRP's continuous enrollment provision to plant up to 500,000 acres of bottomland hardwoods, which are among the most productive U.S. lands for sequestering carbon.

Some programs offer only technical assistance to producers and no financial assistance. USDA has also expanded these programs to encourage GHG emission reductions. For example, the Conservation Technical Assistance (CTA) program lists a reduction in GHG emissions as a national priority. Also, many CTA activities support the scientific underpinnings of the conservation practices that are encouraged by the financial assistance programs, as well as providing the conservation planning requirement for program participation. Compliance programs[42] such as conservation compliance, sodbuster, and swampbuster do not always require specific practices; however, conservation plan requirements under these programs include land management components that could have significant GHG benefits (e.g., tillage requirements, cover crops, and land conversion requirements). Similar to the conservation programs offering financial assistance, technical assistance programs are voluntary and were initiated predominantly for other production or environmental purposes. *Table B-1* highlights conservation practices affecting GHG emissions and USDA programs that offer possible financial or technical assistance for implementation.

USDA recognizes that conservation practices implemented through many of these programs reduce GHG emissions. It estimates that select conservation programs mitigated as much as 68 $MMTCO_2e$ of GHG in 2007 and could potentially mitigate over 81 $MMTCO_2e$ of GHG by 2020 (see *Table 3*). USDA also recognizes that marketable credits may be generated by these conservation programs. Consequently, USDA has recently changed many of its conservation program rules to remove any claim on these credits.[43]

Environmental Services Markets

In addition to expanding several existing conservation programs that were created mainly for purposes other than GHG emission reduction, the 2008 farm bill also included a new conservation provision intended to facilitate the participation of farmers and ranchers in emerging carbon and emissions trading markets. Section 2709 of the bill directed USDA to establish guidelines for standards, accounting procedures, reporting protocols, and verification processes for carbon storage and other types of environmental services markets.[44] This provision was also intended to help address some of the measurement and quantification issues surrounding agricultural and forestry carbon credits, as well as to expand existing voluntary conservation and other farm bill programs, providing incentives that could accelerate opportunities for agriculture and forestry to reduce emissions associated with climate change, adopt energy efficiency measures, and produce renewable energy feedstocks.

Table 3. Estimated GHG Mitigation Impact of USDA Conservation Programs

(Tons of CO_2 equivalent)

	Program 2007	2012	2020
CRP	59,600,000	57,140,000[a]	53,000,000
EQIP	3,938,900	7,877,100	14,178,200
CTA	3,927,600	7,264,000	12,602,300
GRP	7,400	15,400	27,700
WRP	184,000	200,000	250,000
WHIP	251,900	347,800	501,200
CSP	25,400	508,000	685,800
Total	67,935,200	73,352,300	81,245,200

Source: U.S. Department of State, U.S. Climate Action Report 2010, Fifth National Communication of the United States of America Under the United Nations Framework Convention on Climate Change, Washington, DC, June 2010, http://www.state and Richard Swenson, "Conservation Programs that Can be Used to Mitigate Climate Change," Presentation at the Soil and Water Conservation Society Annual Conference, St. Louis, MO, July 20, 2010.

Notes: CRP = Conservation Reserve Program; CSP = Conservation Stewardship Program; CTA = Conservation; Technical Assistance; EQIP = Environmental Quality Incentives Program; GRP = Grassland Reserve Program; WRP = Wetlands Reserve Program; and WHIP = Wildlife Habitat Incentives Program. Estimates of the mitigation impacts of programs are provided by USDA, based on the agency's experience and assumptions related to the implementation of voluntary programs. These estimates may include assumptions about the continued or increased participation of partners, development and deployment goals, and/or whether the necessary commercialization or significant market penetration is achieved.

[a]. Figure represents a 2010 estimate for CRP. A 2012 estimate is not available.

In response to the farm bill provision, USDA created a federal "Conservation and Land Management Environmental Services Board" to assist USDA with the "development of new technical guidelines and science-based methods to assess environmental service benefits which will in turn promote markets for ecosystem services including carbon trading to mitigate climate change."[45] A federally chartered public advisory committee, consisting of farmers, ranchers, forest landowners, and tribal representatives, as well as representatives from state natural resource and agriculture departments, plus public members, was set up to advise the board. USDA also established an Office of Ecosystem Services and Markets to provide administrative and technical assistance in developing the uniform guidelines and tools needed to create and expand markets for ecosystem services in the farming and forestry sectors. Activities of this Office or the Board cannot be identified.[46]

Other USDA Farm Programs

In addition to USDA's conservation programs, several farm bill programs are intended to encourage renewable energy projects and activities that can reduce GHG emissions and/or sequester carbon.

Renewable energy projects received additional program funding in three titles of the 2008 farm bill: Title II (Conservation), Title IX (Energy), and Title VII (Research).[47] One provision in the energy title, the Rural Energy for America Program (REAP), provides mandatory funding for grants for energy audits, renewable energy development, and financial assistance to promote energy efficiency and renewable energy development for farmers and rural small businesses. [48] This program also provides funding to support construction of anaerobic digesters in the livestock sector. Limited information is available regarding the current number of anaerobic digesters[49] installed through REAP. According to a USDA report, since 2003, the Section 9006 grants (precursor to REAP) funded approximately $26 million for anaerobic digesters and have leveraged $123 million in private investment. Over 90 digesters have been funded, of which 19 are operational, 6 are near completion, and 66 are under development.[50]

Renewable energy funding also is available through other federal programs.[51] The 2008 farm bill created the Biomass Crop Assistance Program (BCAP) to assist in the development of renewable energy feedstocks, including cellulosic ethanol, and to provide incentives for producers to harvest, store, and transport biomass.[52] BCAP incentivizes the conversion to dedicated energy crops, which is suggested to help mitigate the negative effects of GHG emissions.[53] Concerns remainder, however, regarding the impact of energy crops on other aspects of the environment (e.g., biodiversity and wildlife habitat).[54] The farm bill's Title VII (Research) also provides for research on renewable fuels, feedstocks, and energy efficiency and for competitive grants for on-farm research and extension projects.

CONCLUSION

Questions of whether and how to address human-induced climate change have been widely debated in recent times. Issues of science, economics, values, geopolitics and a host of other concerns have been contentious. The economic stakes in these debates are potentially large—with both the costs of controls and the "costs of inaction" ranging, by some estimates, into trillions of dollars over several decades.

Results of the 2010 congressional elections have altered political dynamics in Congress on many issues, and leaders of both political parties have indicated that they currently do not plan to pursue comprehensive approaches to addressing climate change in the 112th Congress, although some elements of previous proposals may move through the legislative process—for example, certain energy policy elements. How agriculture fits in these discussions has drawn interest in the past and likely will do so again.

APPENDIX A. AGRICULTURAL SECTOR EMISSIONS OF METHANE AND NITROUS OXIDES

As shown in Table A-1, agricultural activities contributed an estimated 34.2% of all CH_4 emissions and 73.3% of all N_2O emissions in the United States in 2008. Livestock-related categories (enteric fermentation and manure management) were 32.7% of total U.S. CH_4 emissions and 5.4% of total N_2O emissions in 2008, while various land management practices were 8.2% of total U.S. CH_4 emissions and 68% of total N_2O emissions. Between 1990 and 2008, estimated CH_4 emissions from agricultural activities increased by 14.4%, while estimated N_2O emissions fluctuated from year to year, but overall increased by 7.0%.

Table A-1. Estimated Emissions of Methane and Nitrous Oxides: U.S. Agriculture (2008)

Million Metric Tons of CO_2 equivalent (MMTCO_2e)

	MMTCO_2e	Percentage of Total
Total anthropogenic GHG emissions, all sectors	6,956.8	100% of all GHG emissions
Total U.S. methane (CH_4) emissions, all sources	567.6	8.1% of all GHG emissions
Total agriculture CH_4 emissions, all categories	194.0	34.2% of all CH_4 emissions
Enteric Fermentation in Domestic Livestock	140.8	24.8% of all CH_4 emissions
Livestock Manure Management	45.0	7.9% of all CH_4 emissions
Rice Cultivation	7.2	1.3% of all CH_4 emissions
Field Burning of Agricultural Residues	1.0	0.2% of all CH_4 emissions
Total U.S. nitrous oxides (N_2O) emissions, all sources	318.2	4.6 % of all GHG emissions
Total agriculture N_2O emissions, all categories	233.2	73.3% of all N_2O emissions
Agricultural Soil Management	215.9	67.9% of all N_2O emissions
Livestock Manure Management	17.1	5.4% of all N_2O emissions
Field Burning of Agricultural Residues	0.5	0.1% of all N_2O emissions

Source: U,S. Environmental Protection Agency, Inventory of U.S. Greenhouse Gas Emissions and Sinks: 1990-2008, April 15, 2010, EPA 430-R-10-006, Tables 2-1, 6-1; calculations by CRS.

Note: Greenhouse gas emissions also may be measured in teragrams of CO_2 equivalent (Tg CO_2e). One teragram is equal to one million metric tons. The level of certainty for the emissions data varies by source category. Uncertainty rates are more pronounced for the methane and nitrous oxide agricultural source categories due to limited site, crop, and manure management information (see discussion of "Data Uncertainty and Varying Estimates" in the text.)

APPENDIX B. CONSERVATION PRACTICES AND PROGRAMS AFFECTING GHGS

Table B-1. USDA Conservation Programs and Land Management Practices

Conservation Effort	GHG Objectives	General Ecosystem Benefits	USDA Programs
Crops			
Conservation tillage and reduced field pass intensity	Sequestration and emissions	Improves soil, water, and air Reduces soil erosion and fuel quality use	AMA, CB, CCPI, CSP, CTA, EQIP, Compliance, odbuster [a]
Efficient nutrient management	Sequestration and emissions reduction	Improves water quality. Saves expensive, time, and labor	AMA, AWEP, CB, CCPI, CSP, CTA, EQIP
Crop diversity through rotations and cover crops	Sequestration	Reduces erosion and water requirements. Improves soil and water quality	AMA, CB, CCPI, CSP, CTA, EQIP, Compliance, Sodbuster [a]
Livestock			
Manure management	Emissions reduction	On-farm sources of biogas fuel and possibly electricity for large operations. Provides nutrients for crops.	CB, CCPI, CSP, CTA, EQIP, Other [b]
Rotational grazing and improved forage	Sequestration and emissions reduction	Reduces water requirements. Helps withstand drought. Increases long-term grassland productivity	AMA, CCPI, CSP, CTA, EQIP, GLCI,[c] GRP
Feed management through raising feed efficiency and dietary supplements	Emissions reduction	Reduces quantity of nutrients. Improves water quality. Increases efficient use of fuel.	AMA, CB, CCPI, CSP, CTA, EQIP, Other [b]
Agroforestry [d]			
Windbreaks for crops and livestock	Sequestration and emissions reduction	Improves crop and livestock protection and wildlife habitat Provides alternative income sources (e.g., specialty crops and hunting fees).	AMA, CB, CCPI, CRP, CSP, CTA, EQIP, WHIP
Silvopasture with rotational grazing and improved forage Riparian forest buffer	Sequestration and emissions reduction Sequestration	Provides annual income from grazing and long-term income from wood products.	CRP, CSP, CTA, EQIP, GLCI,[c] GRP, WHIP, Other [b]
Riparian forest buffer	Sequestration	Improves water quality and wildlife habitat. Provides alternative	CRP, CSP, CTA, EQIP, GRP, WHIP

Table B-1. (Continued)

Land Use Change			
Conversion of cropland to grassland wetland, or forestland	Sequestration and emissions reduction	Improves soil, water, and air quality. Reduces soil erosion and fuel use. Improves wildlife habitat.	CRP, CTA, GRP, HFRP, WRP, Compliance, Sodbuster, Swampbuster

Source: Compiled by CRS from USDA/NRCS information including USDA, NRCS, Conservation Practices and Programs for Your Land: 2008 Farm Bill, Climate Change Insert, Washington, DC, November 2009.

Notes: Some programs such as CCPI and the Chesapeake Bay Watershed Program (CB) offer additional financial assistance through existing programs like EQIP and WHIP, but are targeted at specific resource concerns or geographic areas.

[a] Conservation compliance, sodbuster, and swampbuster do not always require specific practices; however, conservation plans under these programs could include some of these land management activities.

[b] Renewable energy projects receive additional program funding under titles IX (Energy) and VI (Rural Development) in the 2008 farm bill, as well as other state and federal programs. See CRS Report RL34130, Renewable Energy Programs in the 2008 Farm Bill.

[c] Grazing Lands Conservation Initiative (GLCI) is a USDA initiative that utilizes NRCS technical assistance funds to support private grazing conservation.

[d] Although forestry is not the focus of this report, agroforestry encourages the use of forestry in farming operations to create integrated and sustainable land-use systems. Agroforestry practices are implemented for agriculture production benefits, rather than traditional timber production. Because agroforestry practices can reduce GHG emissions and/or sequester carbon, they are included in this table.

End Notes

[1] This report does not address the underlying debate over global climate change and the role or humans in contributing to it. For discussion on the science and policy of global climate change, see CRS Report RL34266, *Climate Change: Science Highlights*, by Jane A. Leggett, and CRS Report RL34513, *Climate Change: Current Issues and Policy Tools*, by Jane A. Leggett.

[2] Discussion of biofuels is beyond the scope of this report. However, for further information, see CRS Report R41282, *Agriculture-Based Biofuels: Overview and Emerging Issues*, by Randy Schnepf, and CRS Report RL34738, *Cellulosic Biofuels: Analysis of Policy Issues for Congress*, by Kelsi Bracmort et al.

[3] Forestry activities, although not a focus of this report, also can be both a source and a sink of GHGs, releasing GHGs through plant decomposition and removing CO_2 through photosynthesis, storing (sequestering) it in trees, vegetation, and soils.

[4] U.S. Environmental Protection Agency, *Inventory of U.S. Greenhouse Gas Emissions and Sinks: 1990-2008*, April 15, 2010, EPA 430-R-10-006, http://www.epa.gov/climatechange/emissions/usinventoryreport.html. Hereafter, Greenhouse Gas Inventory.

[5] Ibid., p. ES-12.

[6] For further information, see CRS Report RS22964, *Measuring and Monitoring Carbon in the Agricultural and Forestry Sectors*, by Ross W. Gorte and Renée Johnson.

[7] Ibid. Also see CRS Report R40813, *Methane Capture: Options for Greenhouse Gas Emission Reduction*, by Kelsi Bracmort et al.

[8] Different GHGs have varying potential to impact climate change by trapping heat in the atmosphere. In particular, CH_4 and N_2O have greater potency, relative to CO_2. Over a 100-year time horizon, CH_4 emissions have 21 times more potency than equivalent tonnage of CO_2, and N_2O emissions are 310 times more potent than CO_2. To correct for these differences, actual emissions of each GHG are converted to the equivalent amount of CO_2 emissions, based on how potent the substance is compared to CO^2, giving rise to the term "CO_2-equivalent." This measure allows for a comparison of the impacts of emissions of different gases. The figures throughout this report are expressed in carbon equivalents. Greenhouse Gas Inventory, Table ES-1, page ES-3.

[9] See *Table A-1* in the Appendix of this report for more detail.

[10] Enteric fermentation is the production and release of CH4 via eructation (burping) and flatulence as ruminant animals digest their feed.

[11] Lower feed quality and/or higher feed intake leads to higher emissions. Ibid., p. 6-2.

[12] Greenhouse Gas Inventory, p. 6-7.

[13] Greenhouse Gas Inventory., p. 6-27.

[14] For more information, see CRS Report R40874, *Nitrous Oxide from Agricultural Sources: Potential Role in Greenhouse Gas Emission Reduction and Ozone Recovery*, by Kelsi Bracmort.

[15] Greenhouse Gas Inventory, pp. 6-16 – 6-19.

[16] Ibid., pp. 6-6 – 6-7.

[17] Ibid., pp. 6-5, 6-25. To account for data uncertainty and methodological changes in the Inventory, some analysts report emissions averages (e.g., from 2004-2008) rather than single year estimates that are used in this report.

[18] 549 U.S. 497 (2007). For information, see CRS Report RS22665, *The Supreme Court's Climate Change Decision: Massachusetts v. EPA*, by Robert Meltz.

[19] Title V requires major industrial sources of air pollutants to obtain permits which detail all of the federal emission control requirements that apply to the facility.

[20] EPA Administrator Lisa P. Jackson, "Address, 2009 USDA Agricultural Outlook Forum," Feb. 26, 2009, http://www.usda.gov/oce/forum/2009_Speeches/Speeches/Jackson.pdf.

[21] For additional information, see http://www.epa.gov/climatechange/endangerment.html.

[22] For additional information, see CRS Report R40984, *Legal Consequences of EPA's Endangerment Finding for New Motor Vehicle Greenhouse Gas Emissions*, by Robert Meltz. New Source Review (NSR) is a CAA program designed to minimize emissions from new sources and existing sources that are making major modifications.

[23] See "Final GHG Tailoring Rule" at http://www.epa.gov/NSR/actions.html#may10.

[24] Fugitive emissions are generally defined as emissions which are not released, or could not be reasonably collected and released, through a stack or vent.

[25] U.S. Environmental Protection Agency, "Mandatory Reporting of Greenhouse Gases; Proposed Rule," 74 *Federal Register* 16562, April 9, 2009.

[26] U.S. Environmental Protection Agency, "Mandatory Reporting of Greenhouse Gases," 74 *Federal Register* 56260- 56519, October 30, 2009.

[27] U.S. Environmental Protection Agency, Office of Atmospheric Programs, Climate Change Division, *Technical Support Document for Manure Management Systems: Proposed Rule for Mandatory Reporting of Greenhouse Gases*, February 4, 2009, p. 6.

[28] "Climate Bill Slated for House Vote June 26; Waxman, Peterson Announce Deal on Offsets," *Daily Environment Report*, June 26, 2009, p. A-11.

[29] For information, see CRS Report R40643, *Greenhouse Gas Legislation: Summary and Analysis of H.R. 2454 as Passed by the House of Representatives*, coordinated by Mark Holt and Gene Whitney.

[30] For additional information, see CRS Report R40994, *Agriculture and Forestry Provisions in Climate Legislation in the 111th Congress*, by Renée Johnson and CRS Report RL34436, *The Role of Offsets in a Greenhouse Gas Emissions Cap-and-Trade Program: Potential Benefits and Concerns*, by Jonathan L. Ramseur.

[31] Federally supported conservation practices for environmental improvement and agricultural productivity originated with the Soil Conservation and Domestic Allotment Act (P.L. 74-46, 49 Stat. 163) in April 1935, establishing the Soil Conservation Service (SCS) within the USDA. The USDA Natural Resources Conservation Service (NRCS, formally SCS) currently maintains standards for more than 150 conservation practices. A conservation practice standard contains information on why and where the practice is applied, and sets forth the minimum quality criteria that must be met during the application of that practice in order for it to achieve its intended purpose(s).

[32] For example, restoring a wetland would sequester CO_2, but could also release CH_4.

[33] Examples of non-structural practices include crop residue management, nutrient management, and precision agriculture.

[34] As noted previously, there is ongoing debate about the permanence, or duration, of many carbon sequestration practices, depending on the practice itself and such site-specific factors as location, climate and land condition. For information, see CRS Report RS22964, *Measuring and Monitoring Carbon in the Agricultural and Forestry Sectors*, by Ross W. Gorte and Renée Johnson.

[35] For more information on challenges to widespread implementation of anaerobic digestion systems, including the difficulty of quantifying their benefits, see CRS Report R40667, *Anaerobic Digestion: Greenhouse Gas Emission Reduction and Energy Generation*, by Kelsi Bracmort.

36 For information, see CRS Report RS22964, *Measuring and Monitoring Carbon in the Agricultural and Forestry Sectors*, by Ross W. Gorte and Renée Johnson.

37 For additional information on state-level agricultural programs, see CRS Report RL33898, *Climate Change: The Role of the U.S. Agriculture Sector*, by Renée Johnson.

38 Non-industrial private forestland and some forestry activities are eligible under most all conservation programs within USDA. Unlike agroforestry practices, which are used to benefit agriculture production, forestry practices are for the benefit of timber production. While these practices may reduce GHG emissions and increase carbon sequestration, they are not discussed here. For more information on programs that support forestry, see CRS Report RL31065, *Forestry Assistance Programs*, by Ross W. Gorte.

39 For more information on agriculture conservation programs, see CRS Report R40763, *Agricultural Conservation: A Guide to Programs*, by Megan Stubbs.

40 For more information on conservation programs in the 2008 farm bill, see CRS Report RL34557, *Conservation Provisions of the 2008 Farm Bill*, by Tadlock Cowan, Renée Johnson, and Megan Stubbs.

41 For additional discussion on forestry and agricultural activities for carbon sequestration and emissions reduction, see CRS Report RS22964, *Measuring and Monitoring Carbon in the Agricultural and Forestry Sectors*, by Ross W. Gorte and Renée Johnson.

42 Compliance provisions prohibit a producer from receiving many federal farm program benefits (including conservation assistance) when conservation program requirements for highly erodible lands and wetlands are not met.

43 The following program rules include a section recognizing the credits generated by programs and asserting no direct or indirect claim on these credits: EQIP (7 CFR §1466.36), WRP (7 CFR §1467.20), AMA (7 CFR §1465.36), GRP (7 CFR §1415.10), FPP (7 CFR §1491.21), WHIP (7 CFR §636.21), CRP (7 CFR §1410.63(6)), and HFRP (7 CFR §625.8). Also see CRS Report R40692, *Agricultural Conservation Issues in the 111th Congress*, by Megan Stubbs.

44 Section 2709 of P.L. 110-246 amended Section 1245(f) of the Food Security Act of 1985. Ecosystem services refers to the environmental goods and services and other benefits that society obtains from the environment and ecosystems, both natural and managed. Examples include water filtration, flood control, provision of habitat, carbon storage, and many others. For more information, see CRS Report RL34042, *Provisions Supporting Ecosystem Services Markets in U.S. Farm Bill Legislation*, by Renée Johnson.

45 USDA, "USDA Announces New Office of Ecosystem Services and Markets," Release No. 0307.08, Dec. 18, 2008.

46 In March 2010, USDA announced that the Office's title was changed to the Office of Environmental Markets (OEM) and its functions were moved to the USDA Natural Resources and Environment (NRE) mission area. In addition to OEM, USDA's NRE mission area oversees the USDA Forest Service and the Natural Resources Conservation Service.

47 For additional information on renewable energy provisions in the 2008 farm bill, see CRS Report RL34130, *Renewable Energy Programs in the 2008 Farm Bill*, by Megan Stubbs.

48 Previously referred to as Section 9006 (Renewable Energy Systems and Energy Efficiency Improvements) in the 2002 farm bill. Section 9007 of the 2008 farm bill amended the 2002 provision and renamed the program the Rural Energy for America Program.

49 For additional information on anaerobic digesters, see CRS Report R40667, *Anaerobic Digestion: Greenhouse Gas Emission Reduction and Energy Generation*, by Kelsi Bracmort.

[50] William F. Lazarus, Farm-Based Anaerobic Digesters as an Energy and Odor Control Technology: Background and Policy Issues, USDA Office of the Chief Economist Office of Energy Policy and New Uses, Agricultural Economic Report Number 843, February 2008, http://www.usda.gov/oce/reports/energy/AnerobicDigesters0308.pdf, p. 6.

[51] See CRS Report RL34130, *Renewable Energy Programs in the 2008 Farm Bill*, and CRS Report R40110, *Biofuels Incentives: A Summary of Federal Programs*.

[52] For additional information on BCAP, see CRS Report R41296, *Biomass Crop Assistance Program (BCAP): Status and Issues*, by Megan Stubbs.

[53] USDA, FSA, *Biomass Crop Assistance Program Programmatic Environmental Impact Statement*, Final, June 2010.

[54] Les G. Firbank, "Assessing the Ecological Impacts of Bioenergy Projects," *Bioenergy Research*, vol. 1, no. 1 (March 2008), pp. 12-19.

In: Environmental Considerations in Agriculture ISBN: 978-1-61470-077-7
Editors: E.A. Sargsyan, A.J. Ochion 2011 Nova Science Publishers, Inc.

Chapter 3

CLIMATE CHANGE: THE ROLE OF THE U.S. AGRICULTURE SECTOR*

Renée Johnson

SUMMARY

The agriculture sector is a source of greenhouse gas (GHG) emissions, which many scientists agree are contributing to observed climate change. Agriculture is also a "sink" for sequestering carbon, which might offset GHG emissions by capturing and storing carbon in agricultural soils. The two key types of GHG emissions associated with agricultural activities are methane (CH_4) and nitrous oxide (N_2O). Agricultural sources of CH_4 emissions mostly occur as part of the natural digestive process of animals and manure management at livestock operations; sources of N_2O emissions are associated with soil management and fertilizer use on croplands. This report describes these emissions on a carbon-equivalent basis to illustrate agriculture's contribution to total national GHG emissions and to contrast emissions against estimates of sequestered carbon.

Emissions from agricultural activities account for 6%-8% of all GHG emissions in the United States. Carbon captured and stored in U.S. agricultural soils partially offsets these emissions, sequestering about one-tenth of the emissions generated by the agriculture sector, but less than 1% of all U.S. emissions annually. Emissions and sinks discussed in

* This is an edited, reformatted and augmented version of a Congressional Research Service 7-5700, RL33898 publication, dated November 9, 2009.

this report are those associated with agricultural production only. Emissions associated with on-farm energy use or with food processing or distribution, and carbon uptake on forested lands or open areas that might be affiliated with the farming sector, are outside the scope of this report.

Most land management and farm conservation practices can help reduce GHG emissions and/or sequester carbon, including land retirement, conservation tillage, soil management, and manure and animal feed management, among other practices. Many of these practices are encouraged under most existing voluntary federal and state agricultural programs that provide cost-sharing and technical assistance to farmers, predominantly for other production or environmental purposes. However, uncertainties are associated with implementing these types of practices depending on site-specific conditions, the type of practice, how well it is implemented, the length of time a practice is undertaken, and available funding, among other factors. Despite these considerations, the potential to reduce emissions and sequester carbon on agricultural lands is reportedly much greater than current rates.

Congress is currently considering a range of energy and climate policy options. In general, the current climate proposals would not require GHG emission reductions in the agriculture and forestry sectors. However, if enacted, provisions in these bills could potentially raise farm input costs for fossil fuels, fertilizers, energy, and other production inputs. These higher costs could potentially be offset by possible farm revenue increases should farmers participate in carbon offset and renewable energy provisions that are part of this legislation. For example, within capand-trade proposals being debated in Congress are provisions that could provide tradeable allowances to certain agricultural industries, and provisions that could establish a carbon offset program for domestic farm- and land-based carbon storage activities. In addition, the renewable energy provisions contained in these bills could potentially expand the market for farm-based biofuels, biomass residues, and dedicated energy crops. These and related bills and issues are currently being debated in Congress.

INTRODUCTION

The debate in Congress over whether and how to address possible future climate change is intensifying. Often, the role of the U.S. agriculture sector is invoked in this debate. Agriculture is a source of greenhouse gas (GHG) emissions, which many scientists agree are contributing to observed climate change. Agriculture is also a "sink" for sequestering carbon, which partly offsets these emissions. Carbon sequestration (the capture and storage of carbon) in agricultural soils can be an important component of a climate change mitigation strategy, limiting the release of carbon from the soil to the atmosphere.

Congress is considering a range of climate change policy options, including GHG emission reduction programs that would either mandate or authorize a cap-and-trade program to reduce GHG emissions. In general, the current legislative proposals would not require emission reductions in the agriculture and forestry sectors. However, some of these proposals would allow farmers and landowners to generate offsets in support of a cap-and-trade program. Other proposals that Congress has considered would give farmers and landowners a share of available allowances (or credits) for sequestration and/or emission reduction activities. These offsets and allowances could be sold to facilities (e.g., power plants) covered by a cap-and-trade program. Some bills have also specified that the proceeds from auctioned allowances be used to promote certain activities, including farmland conservation and developing bio-energy technologies.

This report is organized in three parts. First, it discusses the extent of GHG emissions associated with the U.S. agriculture sector, and cites current and potential estimates for U.S. agricultural soils to sequester carbon and partly offset national GHG emissions. Second, the report describes the types of land management and farm conservation practices that can reduce GHG emissions and/or sequester carbon in agricultural soils, highlighting those practices that are currently promoted under existing voluntary federal agricultural programs. The *Appendix* provides a summary primer of the key background information presented in these first two sections.

Finally, the report provides a brief overview of legislative action within the ongoing energy and climate debate in Congress to enact changes to existing laws and regulations affecting primarily the energy-producing sectors and other sectors thought to be contributing to GHG concentrations. Many of these energy and climate bills include provisions that could involve farmers and landowners by allowing agriculture and forestry-based carbon offsets and allowances and/or by allowing for farm-based biofuels, biomass residues, and dedicated renewable energy crops. More detail on these bills is available in other CRS reports. This report also describes provisions enacted in the 2008 farm bill (Food, Conservation, and Energy Act of 2008, P.L. 110-246) that could expand the scope of existing farm and forestry conservation programs in ways that could more broadly encompass certain aspects of these climate change initiatives. This farm bill provision is also invoked in most energy and climate bills in order to establish an advisory committee to oversee implementation of agricultural and forestry carbon offsets. The report concludes with a discussion of some of the types of questions that may be raised regarding the role of the U.S. agriculture sector in the broader climate change debate.

This report does not address the potential effects of global climate change on U.S. agricultural production. Such effects may arise because of increased climate variability and incidence of global environmental hazards, such as drought and/or flooding, pests, weeds, and diseases, or temperature and precipitation changes that might cause locational shifts in where and how agricultural crops are produced.[1]

This report also does not address how ongoing or anticipated initiatives to promote U.S. bioenergy production may effect efforts to reduce GHG emissions and/or sequester carbon, such as by promoting more intensive feedstock production and by encouraging fewer crop rotations and planting area setbacks, which could both raise emissions and reduce carbon uptake.[2]

AGRICULTURAL EMISSIONS AND SINKS

Agriculture is a both a source and a sink of greenhouse gases, generating emissions that enter the atmosphere and removing carbon dioxide (CO_2) from the atmosphere through photosynthesis and storing it in vegetation and soils (a process known as sequestration). Sequestration in farmland soils partially offsets agricultural emissions. Despite this offset, however, the U.S. agriculture sector remains a net source of GHG emissions.

Source of National Estimates

Estimates of GHG emissions and sinks for the U.S. agriculture sector presented in this report are the official U.S. estimates of national GHG emissions and carbon uptake, as published annually by the U.S. Environmental Protection Agency (EPA) in its *Inventory of U.S. Greenhouse Gas Emissions and Sinks*.[3] EPA's *Inventory* data reflect annual national emissions by sector and fuel, including estimates for the agriculture and forestry sectors. EPA's estimates rely on data and information from the U.S. Department of Agriculture (USDA), the Department of Energy, the Department of Transportation, the Department of Defense, and other federal departments. The EPA-published data are rigorously and openly peer reviewed through formal interagency and public reviews involving federal, state, and local government agencies, as well as private and international organizations. For the agriculture and forestry sectors, USDA publishes a supplement to EPA's *Inventory*, which builds on much of the same data and information, but in some cases provides a more detailed breakout by individual states and sources.[4]

In this CRS report, emissions from agricultural activities are aggregated in terms of carbon dioxide or CO_2-equivalents, and expressed as million metric tons (MMTCO_2-Eq.).[5] This aggregation is intended to illustrate agriculture's contribution to national GHG emissions and to contrast emissions against estimates of sequestered carbon.

Agricultural Emissions

Total GHG emissions from U.S. agricultural activities have averaged 514 MMTCO_2-Eq. in the past few years (*Table 1*). As a share of total U.S. GHG emissions, the agriculture sector represents about 7% of all estimated annual emissions. Data dating back to 1990 indicate that emissions associated with the U.S. agriculture activities have been increasing, rising from estimated total emissions of 460 MMTCO_2-Eq. in 1990.[6] EPA's reported emissions are expressed in terms of CO_2-equivalent units, and cover both estimated direct emissions and indirect emissions related to electricity use in the sector. These estimates do not cover other types of emissions associated with some agricultural activities, such as carbon monoxide, nitrogen oxides, and volatile organic compounds.

Although the agriculture sector is a leading economic sector contributing to national GHG emissions, its share of total emissions is a distant second compared to that of the energy sector. Fossil fuel combustion is the leading source of GHG emissions in the United States (about 80%), with the energy sector generating 85% of annual emissions across all sectors.[7]

Direct Emissions

The types of direct GHG emissions associated with agricultural activities are methane (CH_4) and nitrous oxide (N_2O), which are among the key gases that contribute to GHG emissions.[8] These gases are significant contributors to atmospheric warming and have a greater effect warming than the same mass of CO_2.[9] Agricultural sources of CH_4 emissions mostly occur as part of the natural digestive process of animals and manure management in U.S. livestock operations.

Table 1. Estimated Current GHG Emissions and Carbon Sequestration: U.S. Agricultural and Forestry Activities, Average 2003-2007

(million metric tons CO_2equiv alent(MMTCO_2 -Eq.))

Source	Emissions	Sequestrationa	Net
Agricultural Activities	513.8	(43.9)	469.9
Direct Emissionsb	484.8	—	—
Indirect electricity-related	29.0	—	—
Land Use Change, Forestryc	31.2	(1,105.2)	(1,074.0)
Subtotal	545.0	(1,149.1)	(604.1)
U.S. Total, All Sources	7,071.2	(1,159.2)	5912.0
% U.S. Total, Agriculture Share	7%	4%	—
% U.S. Total, Forestry Share	<0.5%	95%	—

Source: Compiled by CRS from EPA Inventory of U.S. Greenhouse Gas Emissions and Sinks: 1990-2007, April 2009 (Tables 2-14, 7-3, 7-1, and 6-1), http://epa.gov/climatechange/emissions/usinventoryreport.html. Data shown are five-year average (2003-2007).

Notes:

a) Measured agricultural sequestration categories include land converted to grassland, grassland remaining grassland, land converted to cropland, and cropland remaining cropland. Forestry includes change in forest stocks and carbon uptake from urban trees. Total also includes landfilled yard trimmings and food scraps.

b) Includes CO_2, CH_4, and N_2O. Based on reported emissions attributable to the "agriculture" economic sector, but includes land use and forestry values (EPA *Inventory*, Table 2-14), which are excluded here.

c) Reported as "Emissions from land-use, land-use changes, and forestry" (EPA *Inventory*, Table 7-3).

Sources of N_2O emissions are mostly associated with soil management and commercial fertilizer and manure use on U.S. croplands, as well as production of nitrogen-fixing crops.[10]

Emissions of N_2O from agricultural sources account for about two-thirds of all reported agricultural emissions; emissions of CH_4 account for about one-third of all reported emissions. Across all economic sectors, the U.S. agriculture sector is the leading source of N_2O emissions (about 70%) and a major source of CH_4 emissions (about 25%).[11]

These direct emissions account for the bulk (more than 90%) of estimated emissions associated with U.S. agriculture activities, totaling 530 $MMTCO_2$-Eq. in 2007. Estimates dating back to 1990 indicate that direct emissions from the U.S. agriculture sector have increased steadily, up from about 430 $MMTCO_2$-Eq. in 1990.[12] These estimates do not include emissions associated with on-farm energy use and forestry activities.

Sources of CH_4 and N_2O emissions from agricultural activities are measured across five categories.

- *Agriculture soil management*: Nitrous oxide emissions from farmland soils are associated with cropping practices that disturb soils and increase oxidation, which can release emissions into the atmosphere.[13] The types of practices that contribute to emissions releases are fertilization; irrigation; drainage; cultivation/tillage; shifts in land use; application and/or deposition of livestock manure and other organic materials on cropland, pastures, and rangelands; production of nitrogen-fixing crops and forages; retention of crop residues; and cultivation of soils with high organic content.
- *Enteric fermentation*: Methane emissions from livestock operations occur as part of the normal digestive process in ruminant animals[14] and are produced by rumen fermentation in metabolism and digestion. The extent of such emissions is often associated with the nutritional content and efficiency of feed utilized by the animal.[15] Higher feed effectiveness is associated with lower emissions.
- *Manure management:* Methane and nitrous oxide emissions associated with manure management occur when livestock or poultry manure is stored or treated in systems that promote anaerobic decomposition, such as lagoons, ponds, tanks, or pits.
- *Rice cultivation:* Methane emissions from rice fields occur when fields are flooded and aerobic decomposition of organic material gradually depletes the oxygen in the soil and floodwater, causing anaerobic conditions to develop in the soil, which releases methane.
- *Agricultural residue burning:* Methane and nitrous oxide emissions are released by burning residues or biomass.[16]

The share of GHG emissions for each of these categories is as follows: agriculture soil management (68% of emissions), enteric fermentation (21%), manure management (10%), rice cultivation (1%), and field burning of agricultural residues (less than 1%). Approximately 70% of agricultural emissions are associated with the crop sector and about 30% with the livestock sector (*Figure 1*).[17]

Electricity-Related Emissions

The sector also emits CO_2 and other gases through its on-farm energy use, for example, through the use of tractors and other farm machinery. These emissions are generally aggregated along with other transportation and industrial emissions in the "energy" sources, where they constitute a very small share of the overall total emissions for the sector, estimated at 30 MMTCO_2-Eq.

(*Table 1*). Estimates over the time period since 1990 indicate that emissions associated with electricity use in agriculture activities have been steady or decreasing.[18] These estimates do not include emissions associated with food processing or distribution, which are generally aggregated with emissions for the transportation and industrial sectors.

Land Use and Forestry Emissions

Land use and forestry activities account for less than 1% of total estimated GHG emissions in the United States (*Table 1*). Emissions associated with forestry activities are estimated based on information about forest fires and also land use changes on croplands, wetlands, and peatlands, as well as land conversion and input limitations and management changes.

Uncertainty Estimating Emissions

Agricultural activities may also emit other indirect greenhouse gases, such as carbon monoxide, nitrogen oxides, and volatile organic compounds from field burning of agricultural residues.[19] These emissions are not included in EPA's annual *Inventory* estimates because they contribute only indirectly to climate change by influencing tropospheric ozone, which is a greenhouse gas. Agricultural activities may also release other types of air emissions, some of which are regulated under the federal Clean Air Act, including ammonia, volatile organic compounds, hydrogen sulfide, and particulate matter.[20] These types of emissions are typically not included in proposals to limit GHG emissions.

EPA's estimates are based on annual USDA data on crop production, livestock inventories, and information on conservation and land management practices in the agriculture sector.

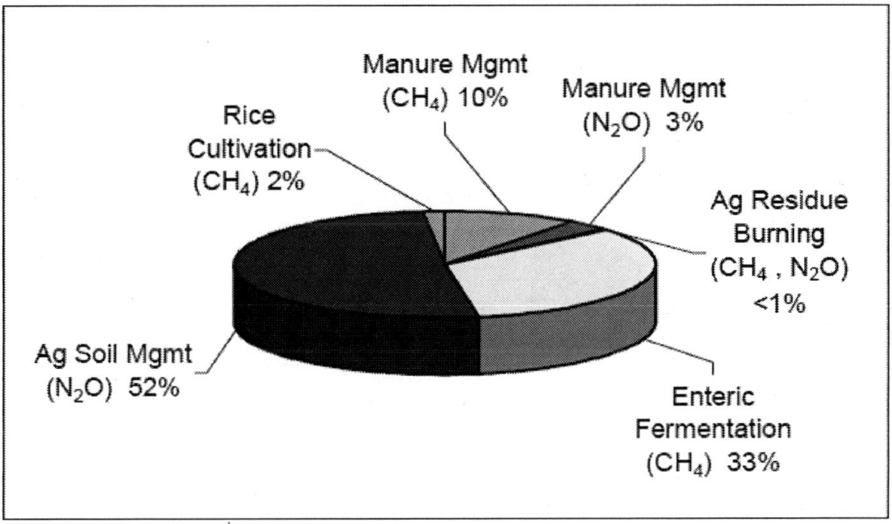

Source: EPA Inventory.

Figure 1. Agricultural GHG Emissions, Average 2003-2007.

Actual emissions will depend on site-specific factors, including location, climate, soil type, type of crop or vegetation, planting area, fertilizer and chemical application, tillage practices, crop rotations and cover crops, livestock type and average weight, feed mix and amount consumed, waste management practices (e.g., lagoon, slurry, pit, and drylot systems), and overall farm management. Emissions may vary from year to year depending on actual growing conditions. The EPA-reported data reflect the most recent data and historical updates, and reflect underlying methodological changes, in keeping with Intergovernmental Panel on Climate Change (IPCC) guidelines.[21] More detailed information is in EPA's *Inventory*.

Potential for Additional Emission Reductions

There is potential to lower GHG emissions from U.S. agricultural facilities at both crop and livestock operations through further adoption of certain conservation and land management practices. In most cases, such practices may both reduce emissions and sequester carbon in agricultural soils.

Improved Soil Management

Options to reduce nitrous oxide emissions associated with crop production include improved soil management, more efficient fertilization, and implementing soil erosion controls and conservation practices. In the past 100 years, intensive agriculture has caused a soil carbon loss of 30%-50%, mostly through traditional tillage practices.[22] In contrast, conservation tillage practices preserve soil carbon by maintaining a ground cover after planting and by reducing soil disturbance compared with traditional cultivation, thereby reducing soil loss and energy use while maintaining crop yields and quality. Practices include no-till and minimum, mulch, and ridge tillage. Such tillage practices reduce soil disturbance, which reduces oxidation and the release of carbon into the atmosphere. Therefore, conservation tillage practices reduce emissions from cultivation and also enhance carbon sequestration in soils (discussed later in this report). Nearly 40% of U.S. planted areas are under some type of conservation tillage practices.[23]

Improved Manure and Feed Management

Methane emissions associated with livestock production can be reduced through improved manure and feed management. Improved manure management is mostly associated with installing certain manure management systems and technologies that trap emissions, such as an anaerobic digester[24] or lagoon covers. Installing such systems generates other principal environmental benefits. Installing an anaerobic digester to capture emissions from livestock operations, for example, would also trap other types of air emissions, including air pollutants such as ammonia, volatile organic compounds, hydrogen sulfide, nitrogen oxides, and particulate matter that are regulated under the federal Clean Air Act. Other benefits include improved water quality through reduced nutrient runoff from farmlands, which may be regulated under the federal Clean Water Act.[25] Many manure management systems also control flies, produce energy, increase the fertilizer value of any remaining biosolids, and destroy pathogens and weed seeds.[26]

Manure management systems, however, can be costly and difficult to maintain, given the typically high start-up costs and high annual operating costs. For example, the initial capital cost of an anaerobic digester with energy recovery is between $0.5 million and $1 million at a large-sized dairy operation, and annual operating costs are about $36,000. Initial capital costs for a digester at a larger hog operation is about $250,000, with similar operating costs.[27] Upfront capital costs tend to be high because of site-specific conditions at an individual facility, requiring technical and engineering expertise. Costs will vary depending on site-specific conditions but may also vary by production region. Costs may be higher in areas with colder temperatures, where some types of digesters may not be appropriate or may require an additional heat source, insulation, or energy requirements to maintain constant, elevated temperatures.[28] Energy requirements to keep a digester heated are likely be lower in warmer climates.

Incentives are available to assist crop and livestock producers in implementing practices and installing systems that may reduce GHG emissions. Such incentives include cost-sharing and also low-interest financing, loan guarantees, and grants, as well as technical assistance with implementation. Funding for anaerobic digesters at U.S. livestock operations has been available to livestock producers under various farm bill programs.[29] Despite the availability of federal and/or state-level cost-sharing and technical assistance, adoption of such systems remains low throughout the United States. There are currently about 100 digester systems in operation or planned at commercial dairy and hog farms, accounting for only 1% of operations nationwide (*Figure 2*).[30]

Improved feed strategies may also lower methane emissions at livestock operations. Such strategies may involve adding supplements and nutrients to animal diets, substituting forage crops for purchased feed grains, or instituting multi-phase feeding to improve digestive efficiency. Other options involve engineering genetic improvements in animals.[31] Purchasing feed supplements and more intensely managing animal nutrition and feeding practices may add additional costs and management requirements at the farm level.

AGRICULTURAL CARBON SINKS

Carbon Loss and Uptake

Agriculture can sequester carbon, which may offset GHG emissions by capturing and storing carbon in agricultural soils. On agricultural lands, carbon can enter the soil through roots, litter, harvest residues, and animal manure, and may be stored primarily as soil organic matter (SOM; see *Figure 3*).[32] Soils can hold carbon both underground in the root structure and near the soil surface and in plant biomass. Loss of soil carbon may occur with shifts in land use, with conventional cultivation (which may increase oxidation), and through soil erosion. Carbon sequestration in agricultural soils can be an important component of a climate change mitigation strategy, since the capture and storage of carbon may limit the release of carbon from the soil to the atmosphere.

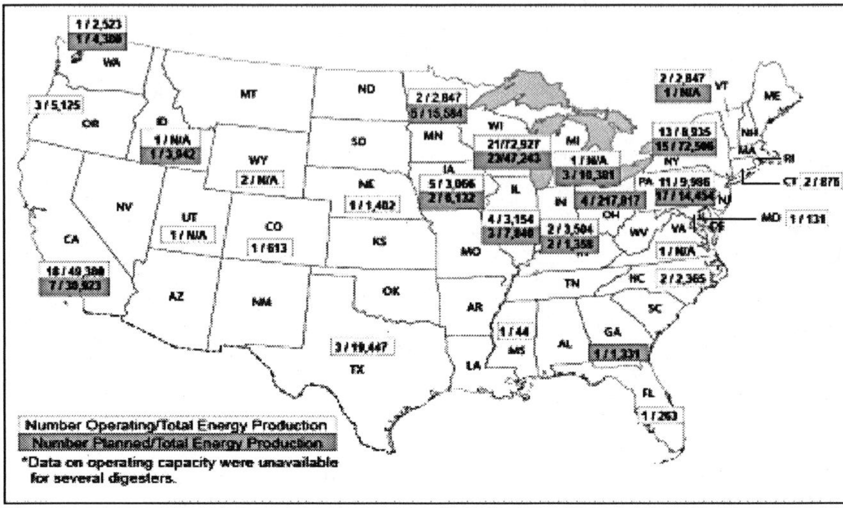

Source: Adapted by CRS, Map Resources (7/2007) from data reported by USEPA, AgStar Digest, Winter 2006.

Figure 2. National Distribution of Anaerobic Digester Energy Production.

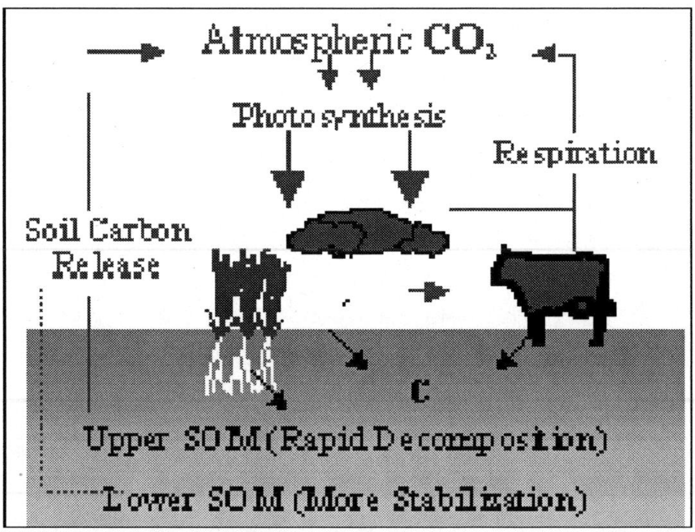

Source: USGS, "Carbon Sequestration in Soils."

Note: SOM = Soil organic matter.

Figure 3. Carbon Sequestration in Agricultural Soils.

Voluntary land retirement programs and programs that convert or restore grasslands and wetlands promote carbon capture and storage in agricultural soils. Related practices include afforestation (including the conversion of pastureland and cropland), reforestation, and agro-forestry practices. Conservation practices that raise biomass retention in soils and/or reduce soil disturbance, such as conservation tillage and/or installing windbreaks and buffers, also promote sequestration. More information is provided in the report section "Conservation Practices that Promote Mitigation ."

CRS Report RS22964, *Measuring and Monitoring Carbon in the Agricultural and Forestry Sectors*, summarizes estimated sequestration rates for selected types of farm and forestry practices, based on the current literature as summarized by USDA and EPA.[33]

Agriculture-Based Sequestration

Total carbon sequestration from U.S. agricultural activities has averaged about 44 $MMTCO_2$-Eq. during the 2003-2007 time period (*Table 1*). Compared to total agriculture-based emissions, sequestration within the sector accounts for only a small share (less than 10%) of its annual emissions. Compared to total U.S. GHG emissions, agriculture-based sequestration accounts for less than 1% of emissions each year.[34] Data dating back to 1990 indicate that carbon sequestration associated with U.S. agriculture activities has decreased significantly, from an estimated total storage of 96 $MMTCO_2$-Eq. in 1990 to 45 $MMTCO_2$-Eq. in 2007.[35] Carbon sequestration in the U.S. agriculture sector currently offsets only about 5% of the carbon-equivalent of reported GHG emissions generated by the agriculture sector each year. Thus the sector remains a net source of GHG emissions.

Other Land Use and Forestry Sequestration

These estimates do not include estimates for the forestry sector, or sequestration activities on forested lands or open areas that may be affiliated with the agriculture sector. Forests and trees account for a majority (about 95%) of all estimated carbon uptake in the United States, mostly through forest restoration and tree-planting. As shown in *Table 1*, land use and forestry practices account for a much larger share of annual carbon storage from land-based systems, and are estimated to have averaged 1,105 $MMTCO_2$-Eq. during the past few years. Compared to total U.S. GHG emissions, sequestration from land use and forestry practices accounts for about 16% of emissions each year. Historical data show that carbon sequestration from land use and forestry activities has increased, rising from an estimated storage of 660 $MMTCO_2$-Eq. in 1990 to 910 $MMTCO_2$-Eq. in 2007.[36]

The agriculture and forestry sectors are only part of the overall carbon sequestration debate. Carbon sequestration by these sectors is usually referred to as indirect or biological sequestration.[37] Biological sequestration is considered to have less potential for carbon sequestration than direct sequestration, also referred to as carbon capture and storage, and is typically associated with oil and gas production.

Uncertainty Estimating Carbon Sinks

EPA's *Inventory* estimates of carbon uptake in agricultural soils are based on annual data and information on cropland conversion to permanent pastures and grasslands, reduced summer fallow areas in semi-dry areas, increased conservation tillage, and increased organic fertilizer use (e.g, manure) on farmlands, as well as information on adoption rates and use of certain conservation and land management practices.

However, actual carbon uptake in agricultural soils depends on several site-specific factors, including location, climate, land history, soil type, type of crop or vegetation, planting area, tillage practices, crop rotations and cover crops, and farm management in implementing certain conservation and land management practices. Estimates of the amount of carbon sequestered may vary depending on the amount of site-specific information included in the estimate, as well as on the accounting procedures and methodology used to make such calculations.

In general, the effectiveness of adopting conservation and land management practices will depend on the type of practice, how well the practice is implemented, and also on the length of time a practice is undertaken. For example, time is needed for a certain conservation practice to take hold and for benefits to accrue, such as buildup of carbon in soils from implementing conservation tillage or other soil management techniques, and growing time for cover crops or vegetative buffers. The overall length of time the practice remains in place is critical, especially regarding the sequestration benefits that accrue over the time period in which land is retired. In addition, not all conservation and land management practices are equally effective or appropriate in all types of physical settings. For example, the use and effectiveness of conservation tillage practices will vary depending on soil type and moisture regime, which may discourage some farmers from adopting or continuing this practice in some areas.

The potential impermanence of conservation and land management practices raises concerns about the effectiveness and limited storage value of the types of conservation practices that sequester carbon, given that the amount of carbon stored depends on the willingness of landowners to adopt or continue to implement a particular voluntary conservation practice. There are also concerns that the addition of other conservation practices may not significantly enhance the sequestration potential of practices that might already be in place.[38] This raises questions about the cost-effectiveness of sequestering carbon on farmlands relative to other climate change mitigation strategies in other industry sectors. Finally, implementing conservation practices and installing new technologies may be contingent on continued cost-sharing and other financial incentives contained in the current farm bill; programs funded through this legislation help offset the cost to farmers for these practices and technologies, which some farmers may not be willing to do otherwise.

Potential for Additional Uptake

USDA reports that the potential for carbon uptake in agricultural soils is much greater than current rates. USDA forecasts that the amount of carbon sequestered on U.S. agricultural lands will more than double from current levels by 2012, adding roughly an additional 40 MMTCO_2- Eq. of sequestered carbon attributable to the sector.[39] This additional uptake is expected through improved soil management (roughly 60%), improved manure and nutrient management (about 30%), and additional land-retirement sign-ups (about 10%). Longer-term estimates from USDA and EPA report that the potential for net increases in carbon sequestration in the agriculture sector could reach an estimated 590 to 990 MMTCO_2-Eq. per year (*Table 2).*

An additional carbon uptake potential of 590 to 990 MMTCO_2-Eq. per year would more than offset the agriculture sector's annual GHG emissions, or offset 8% to 14% of total current national emissions from all sources. Currently, carbon uptake in agricultural soils sequesters under 1% of total national GHG emissions annually (*Table 1*). Many U.S. farm groups claim that the U.S. agriculture sector has the potential to store between 15% and 25% of total annual U.S. emissions,[40] but it is unclear whether this cited potential also includes already substantial sequestration from current land use and forestry practices. An estimated 16% of all GHG emissions are currently sequestered annually, with the bulk through growth in forest stocks.

Studies by both the USDA and EPA provide aggregate annual estimates of the additional carbon storage potential for various agricultural and forestry activities (*Table 2*). These aggregate estimates are in addition to current estimated sequestration rates in these sectors (*Table 1*).

The USDA and EPA studies both account for current conditions, as well as expected direct costs and opportunity costs in modeling landowners' decision-making. These estimates are measured in terms of carbon storage over time (15 to 100 years) across a range of assumed carbon market prices (roughly \$3 to \$50/MT CO_2-Eq.). These published results show a range of carbon prices by type of farming and forestry activity. The presumed relationship between carbon sequestration and price shows that as carbon prices rise, this will likely attract more investment and adoption of additional and differing types of mitigation activities. These estimates are reported as a national total and are also broken out by select U.S. regions.

Table 2 shows the estimated carbon mitigation potential reported by EPA and USDA for two mitigation categories—afforestation and soil sequestration—across a range of assumed carbon prices. In general, the low end of this price range indicates that carbon sequestration potential is mostly associated with cropland management practices, whereas higher-end prices are mostly associated with land retirement and conversion, and a longer sequestration tenure. EPA's analysis includes estimates of other mitigation activities, including forest management on private lands. These estimates reflect the net reduction compared to baseline conditions, or current estimated sequestration (*Table 1*).

USDA reports that the potential for net increases in carbon sequestration through afforestation and in agricultural soils is estimated to range widely from 0 to 587 MMT CO_2-Eq. per year, following the implementation of a 15-year program (*Table 2*).[41] Sequestration potential is estimated to be greatest at the high end of the assumed price range for carbon (about \$30/MT CO_2-Eq.). At this price level, USDA projects sequestration levels could increase by 587 MMT CO_2-Eq. annually. Even at lower prices (about \$3/MT CO_2-Eq.), the projected mitigation potential is double the current estimated sequestration for these types of agricultural activities. Comparable EPA estimates (15-year period) project a higher sequestration potential for the U.S. agricultural sector across the range of assumed carbon prices, reported at 160 MMT CO_2-Eq. per year at lower carbon prices to 990 MMTCO_2-Eq. per year at the higher price levels.[42]

Table 2. Carbon Sequestration Potential in the U.S. Agriculture Sector, Alternative Scenarios and Payment Levels

(dollars per million metric ton of sequestered CO_2)

Source	\$3-\$5 range	\$14-\$15 range	\$30-\$34 range
USDA Estimate	(million mt of sequestered CO_2)		
Afforestation	0-31	105-264	224-489
Agricultural soil carbon sequestration	0.4-4	3-30	13-95
Total	0.4-35	108-295	237-587
EPA Estimate			
Afforestation	12	228	806
Agricultural soil carbon sequestration	149	204	187
Total	161	432	994

Sources: EPA, Greenhouse Gas Mitigation Potential in U.S. Forestry and Agriculture, Nov. 2005, Table 4-10, at http://www.epa.gov/sequestration/greenhouse. Compares USDA estimates (Economics of Sequestering Carbon in the U.S. Agricultural Sector, Apr. 2004) with EPA estimates.

For information on USDA and EPA estimates and how these estimates were derived, see CRS Report RS22964, *Measuring and Monitoring Carbon in the Agricultural and Forestry Sectors.*

Afforestation (creation of forested areas mostly through conversion of pastureland and cropland) reflects the majority of the estimated uptake potential, with agricultural soil carbon sequestration accounting for a smaller share at the high end of the estimated range. However, large projected gains in mitigation from afforestation could be overly optimistic, given that afforestation is highly dependent on land availability and may only come from available cropland or pastureland. However, as reported by the Congressional Budget Office (CBO), estimates of the future mitigation potential from afforestation and cropland soil sequestration often vary significantly across different studies.[43]

In March 2009, EPA indicated that it had updated its underlying model and subsequently its estimates of the carbon mitigation potential from farm and forestry practices.[44] Underlying changes to EPA's simulation models are reflected in EPA's June 2009 analysis of the House-passed climate bill, H.R. 2454, which includes an analysis of the estimated effects of the bill's carbon offset program for certain mitigation activities on agriculture and forest lands.[45] EPA's current analysis predicts that the mitigation potential from agriculture soil carbon activities will be largely outweighed by other types of mitigation activities, including forest, manure, and crop management, which are now predicted to account for a greater share of overall mitigation potential compared to previous EPA estimates. For more information about EPA's model and estimates, see CRS Report R40236, *Estimates of Carbon Mitigation Potential from Agricultural and Forestry Activities.*

Enhancing Carbon Sinks

There is potential to increase the amount of carbon captured and stored in U.S. agricultural lands by adopting certain conservation and land management practices. In most cases, such practices may both sequester carbon in farmland soils and reduce emissions from the source.

Estimates of representative carbon sequestration rates for selected types of farm and forestry practices are provides in CRS Report RS22964, *Measuring and Monitoring Carbon in the Agricultural and Forestry Sectors.*

Improved Soil and Land Management

The main carbon sinks in the agriculture sector are cropland conversion and soil management, including improved manure application.[46] More than half of all carbon sequestered on U.S. agricultural lands is through voluntary land retirement programs and programs that convert or restore land (e.g., conversion to open land or grasslands, conversion to cropland, restoration of grasslands or wetlands, etc.). Undisturbed open lands, grasslands and wetlands can hold carbon in the soil both underground in the root structure and above ground in plant biomass. The amount of carbon sequestered will vary by the type of land management system. Afforestation and cropland conversion have the greatest potential to store the most carbon per acre annually, compared with other types of systems, such as tree plantings and wetlands conversion, or storage in croplands.[47]

Conservation tillage is another major source of sequestration on farmlands, accounting for about 40% of the carbon sequestered by the U.S. agriculture sector. [48] Improved tillage practices improve biomass retention in soils and reduce soil disturbance, thereby decreasing oxidation. The amount of carbon sequestered will vary by the type of tillage system. Among conservation tillage practices, no-till stores about 30% more than the amount of carbon stored by reduced tillage but more than five times that stored on intensive tilled croplands. (Conservation tillage practices are explained in the section on "Potential for Additional Emission Reductions.")

Improved Manure and Feed Management

Mitigation strategies at U.S. livestock operations are not commonly associated with carbon uptake and are not included in EPA's carbon sink estimates. However, installing manure management systems, such as an anaerobic digester, captures and/or destroys methane emissions from livestock operations and may be regarded as avoided emissions or as a form of direct sequestration capturing emissions at the source. As a result, many carbon offset programs are promoting manure management systems as a means to capture and store methane at dairy operations, which may also be sold as carbon offset credits and as a renewable energy source.[49] Given that there are currently few anaerobic digesters in operation, estimates of the actual or potential uptake may be difficult to estimate. (Manure management systems are further explained in the section on "Potential for Additional Emission Reductions.")

Improved dietary and feed management strategies may also lower methane emissions by reducing intestinal methane in livestock. Research in this area is still ongoing. As already noted, such strategies may involve adding supplements and nutrients to animal diets, substituting forage crops for purchased feed grains, or instituting multiphase feeding to improve digestive efficiency. Some noted strategies include feeding cattle flaxseed, alfalfa, and grasses high in Omega-3 fatty acids, and managing animal nutrition and feeding practices. Genetic improvements in animals might also lower intestinal methane. Guidelines will likely vary depending on location, nutritional requirements, management strategy, and animal type.

Conservation Practices that Promote Mitigation

Existing conservation and farmland management programs administered at both the federal and state levels often encourage the types of agricultural practices that can reduce GHG emissions and/or sequester carbon. These include conservation, forestry, energy, and research programs within existing farm legislation. These programs were initiated predominantly for other production or environmental purposes, and few specifically address climate change concerns in the agriculture and forestry sectors. However, some USDA and state-level programs have started to place additional attention on the potential for emissions reduction and carbon storage under certain existing programs.

Agricultural conservation and other farmland practices broadly include land management, vegetation, and structures that can also reduce GHG emissions and/or sequester carbon, such as:

- *Land retirement, conversion, and restoration*—conversion/restoration to grasslands, wetlands, or rangelands; and selected structural barriers, such as vegetative and riparian buffers, setbacks, windbreaks;
- *Cropland tillage practices*—reduced/medium- till, no-till, ridge/strip-till vs. conventional tillage;
- *Soil management/conservation*—soil supplements/amendments, soil erosion controls; precision agriculture practices, recognized best management practices;
- *Cropping techniques*—crop rotations, cover cropping, precision agriculture practices, efficient fertilizer/nutrient (including manure) and chemical application;
- *Manure and feed management*—improved manure storage (e.g., anaerobic digestion, methane recovery); and improved feed efficiency, dietary supplements;
- *Grazing management*—rotational grazing, improved forage practices; *Bioenergy/biofuels substitution*—on-farm use, replacing fossil fuels or deriving bioenergy from land-based feedstocks, renewable energy; and
- *Energy efficiency and energy conservation* (on-farm).

In general, conservation programs administered by USDA and state agencies encourage farmers to implement certain farming practices and often provide financial incentives and technical assistance to support adoption. Participation in these programs is voluntary, and farmers may choose to discontinue participating in these programs. The effectiveness of these practices depends on the type of practice, how well the practice is implemented, and also on the length of time a practice is undertaken. These programs are generally designed to address site-specific improvements based on a conservation plan developed with the assistance of USDA or state extension technical and field staff that considers the goals and land resource base for an individual farmer or landowner. Such a conservation plan is typically a necessary precursor to participating in USDA's conservation programs.

Although not the focus of this report, forestry practices that reduce emissions and/or sequester carbon include afforestation and reforestation; forest management (such as harvest for long-term wood products, reduced-impact logging, certified sustainable forestry, thinning/release, and fertilization); pruning; and avoided deforestation and forest degradation.[50]

Federal Programs

Conservation Programs

Conservation programs administered by USDA are designed to take land out of production and to improve land management practices on land in production, commonly referred to as "working lands" (*Table 3*). These programs are provided for in Title II (Conservation) of the 2008 farm bill (P.L. 110-246, the Food, Conservation, and Energy Act of 2008).

- *Land retirement/easement programs.* Programs focused on land management, including programs that retire farmland from crop production and convert it back into forests, grasslands, or wetlands, including rental payments and cost-sharing to establish longer term conservation coverage. Major programs include the Conservation Reserve Program (CRP), the Wetlands Reserve Program (WRP), the Grasslands Reserve Program (GRP), the Farmland Protection Program (FPP).
- *Working lands programs.* Programs focused on improved land management and farm production practices, such as changing cropping systems or tillage management practices, are supported by cost-sharing and incentive payments, as well as technical assistance. Major programs include the Environmental Quality Incentives Program (EQIP), the Conservation Stewardship Program (CSP), the Agricultural Management Assistance (AMA) program, and the Wildlife Habitat Incentives Program (WHIP).

Prior to the 2008 farm bill, few USDA conservation programs were specifically intended to address climate change concerns in the agriculture sector. One exception is USDA's Conservation Innovation Grants program, a subprogram under EQIP that provides for competitive awards, and is intended to accelerate technology transfer and adoption of innovative conservation technologies, mostly through pilot projects and field trials. Past grants have supported development of approaches to reduce ammonia emissions from poultry litter, promote conservation tillage and solar energy technologies, and develop private carbon sequestration trading credits.[51]

USDA has expanded some of its existing farmland conservation programs to further encourage emission reductions and carbon sequestration. Many of the practices encouraged under EQIP and CSP reduce net emissions. USDA has provided additional technical guidance to make GHG a priority resource concern in EQIP and CSP by giving greater weight to projects that promote anaerobic digestion, nutrient management plans, and other types of cropland practices, such as installing shelter belts and windbreaks, encouraging conservation tillage, and providing resources for biomass energy projects. Programs such as CTA, AMA, EQIP, and CSP list a reduction in emissions as a national priority for the program, which effects the funding and ranking of projects. Under CRP, USDA has modified how it scores and ranks offers to enroll land in CRP in order to place greater weight on installing vegetative covers that sequester carbon. USDA also has an initiative under CRP's continuous enrollment provision to plant up to 500,000 acres of bottomland hardwoods, which are among the most productive U.S. lands for sequestering carbon. As of April 2009, more than 45,000 acres have been enrolled in this initiative.

In addition, USDA has recognized that marketable credits may be generated by these conservation programs and has removed any claim on these credits through recent changes to many of its conservation program rules.[52]

Not including funding increases authorized under the 2008 farm bill, actual funding for USDA's conservation programs has totaled more than $5 billion annually. Voluntary land retirement programs and programs that convert or restore land account for roughly 37% annually of all USDA conservation spending (*Figure 4*). Programs that provide cost-sharing and technical assistance to farmers to implement certain practices, such as EQIP, CSP, and AMA, provide another 21% annually. USDA's conservation technical assistance and extension services account for about one-fourth of all funding. Other federal funding through other programs also generally promotes natural resource protection on U.S. farms. Generally, the decision on how and where this funding is ultimately used is made at the individual state level.

Table 3.Conservation and Land Management Practices

USDA Program	Conservation Practice and Land Management	General Benefits	Benefits for Climate Change
EQIP CSP AMA	Conservation tillage and reduced field pass intensity	Improves soil/water/air quality. Reduces soil erosion/fuel use.	Sequestration, emission reduction
	Crop diversity through crop rotations and cover cropping	Reduces erosion/water needs. Improves water quality	Sequestration
	Efficient nutrient (nitrogen) management, fertilizer application	Saves expenses, time, and labor	Sequestration, emission reduction
	Improved soil management and soil erosion controls	Improves soil/water/air quality	Sequestration, emission reduction
EQIP CSP AMA Other [a]	Manure management (e.g., storage/containment, anaerobic digestion and methane recovery)	On-farm fuel cost-savings. Alternative income source. Nutrients for crops.	Emission reduction
EQIP CSP AMA	Feed management (e.g., raise feed efficiency, dietary supplements)	Improves water/air quality. More efficient use of feed	Emission reduction
	Rangeland management (e.g., rotational grazing, improved forage)	Reduces water requirements Helps withstand drought Raises grassland. productivity.	Sequestration, emission reduction
EQIP CSP AMA WHIP	Windbreaks for crops and livestock, vegetative/riparian buffers, grassed waterways, setbacks, etc.	protection and wildlife habitat. Alternative income source (e.g., hunting fees).	Sequestration, emission reduction
EQIP CSP AMA	Agroforestry/silvopasture with rotational grazing and improved forage	Provides income from grazing and wood products.	Sequestration, emission reduction

Table 3. (Continued)

USDA Program	Conservation Practice and Land Management	General Benefits	Benefits for Climate Change
CRP WRP GRP FPP	Land management, including retirement, conversion, restoration (cropland, grasslands, wetlands, open space)	Improves soil/water/air quality.	Sequestration
EQIP CSP AMA Other [a]	Energy efficiency/conservation	Improves soil/water/air quality. Cost-savings.	Emission reduction
	Biofuel substitution and renewable energy use	Improves soil/water/air quality. On-farm fuel cost-savings. Alternative income source.	Emission reduction

Reserve Program (CRP), Wetlands Reserve Program (WRP), Grasslands Reserve Program (GRP), Farmland Protection Program (FPP), Environmental Quality Incentives Program (EQIP), Conservation Stewardship Program (CSP), Agricultural Management Assistance (AMA), Wildlife Habitat Incentives Program (WHIP).

[a]. Renewable energy projects receive additional program funding in farm bill under Title IX (Energy) and Title VI (Rural Development), as well as other federal and state program.

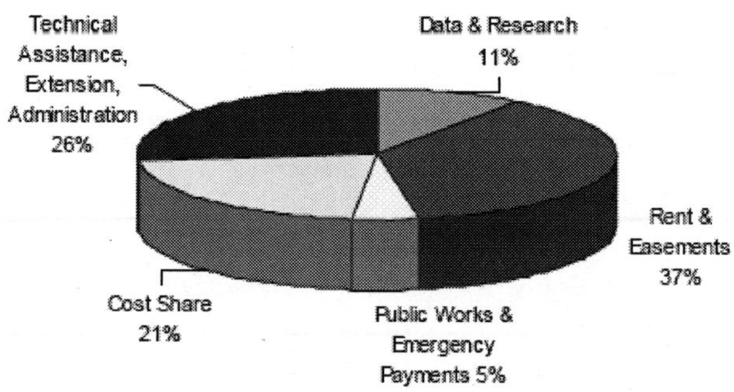

Source: USDA, Office of Budget and Planning.
Note: FY2005 total spending = $5.6 billion.

Figure 4. USDA Conservation Spending, FY2005.

The 2008 farm bill expanded mandatory funding for several existing conservation programs that contribute to increased carbon storage in soil and plants, reduced agriculture-based emissions associated with climate change, lowered energy consumption by farming operations, and increased production of renewable fuels and feedstocks, among other provisions.

In particular, the 2008 farm bill increased funding for both EQIP and CSP, and expanded eligibility to include management practices on private forest lands and other natural resource areas. The farm bill also provided funding for the Conservation Innovation Grants program to address air quality concerns from agriculture operations, including greenhouse gas emissions. It also made changes to USDA's land retirement programs. Changes to CRP are expected to encourage the establishment of native vegetation cover on lands set aside or retired from agricultural production, and promote tree planting and management to improve habitat and encourage healthy forest growth and carbon uptake. Changes to FPP include expanded eligibility for forest lands, and changes to GRP include expanded grasslands enrollment and emphasis on long-term and permanent easement. The farm bill also included a new conservation provision intended to facilitate the participation of farmers and ranchers in emerging carbon and emissions trading markets by directing USDA to establish guidelines for standards, accounting procedures, reporting protocols, and verification processes for carbon storage and other types of environmental services markets. (This new provision is described in further detail in the section on "2008 Farm Bill Provisions").

Other Farm Programs

Aside from USDA's conservation programs, there are other farm bill programs that encourage the types of agricultural practices that can reduce GHG emissions and/or sequester carbon. These include programs in the farm bill's forestry, energy, and research titles.[53]

Renewable energy projects receive additional program funding across three farm bill titles: Title II (Conservation), Title IX (Energy), and Title VII (Research). In addition to cost-sharing provided under USDA's conservation programs, one energy title provision in the 2008 farm bill is the Rural Energy for America Program (Section 9007). This program provided mandatory funding for grants for energy audits, renewable energy development, and financial assistance to promote energy efficiency and renewable energy development for farmers and rural small businesses.[54] In the past this program has provided funding to support construction of anaerobic digesters in the livestock sector.[55] Other renewable energy funding is also available through other federal programs.[56] The 2008 farm bill also created the Biomass Crop Assistance Program to assist in the development of renewable energy feedstocks, including cellulosic ethanol, and to provide incentives for producers to harvest, store, and transport biomass. The farm bill's Title VII (Research) also provided for research on renewable fuels, feedstocks, and energy efficiency and for competitive grants for on-farm research and extension projects.

Forestry programs, administered by USDA's Forest Service, are provided for in Title VIII (Forestry) of the farm bill. Typically, there is often little overlap between the various agriculture and forestry programs administered by USDA, and few forestry programs provide support to agricultural enterprises.[57] One program with an agroforestry component is the Healthy Forests Reserve Program, which was reauthorized in the 2008 farm bill. This program assists with restoring and enhancing forest ecosystems; however, funding for this program is usually limited to a few states. The 2008 farm bill also created new programs with possible agroforestry benefits, including (1) the Community Forest and Open Space Conservation Program, authorizing new cost-share grants for local governments, tribes, and non-profits to acquire lands threatened by conversion to non-forest uses; and (2) the Emergency Forest Restoration Program, providing for the rehabilitation of croplands, grasslands, and private non-industrial forests following natural disasters. The farm bill also expanded or created other programs to protect and restore privately owned forests, which could also contribute to retaining or increasing carbon storage capacity on forest lands.

State Programs

Agriculture Conservation and Land Management Programs

State-level agriculture conservation and land management programs are available to farmers in most states, and operate in much the same manner as federal conservation programs. These programs may also provide financial and technical assistance to farmers to implement certain practices, using additional state resources and in consultation with state agriculture agencies and extension staff. No single current compendium exists outlining the different types of agriculture conservation programs across all states; instead information is available through individual state government websites.[58]

Many states have cost-share programs that provide financial assistance to landowners to implement practices that benefit a state's forests, fish, and wildlife. Many of these programs provide technical assistance and up to 75% of the eligible costs of approved conservation projects to qualified landowners. Several states also provide low-interest financing to farmers and landowners to encourage conservation practices or to implement best management practices for the agriculture sector. Many states also have buffer strip programs, which may provide rental payments to landowners who agree to create or maintain vegetative buffer strips on croplands near rivers, streams, ponds, and wetlands. Typically states that have taxing authority for conservation purposes, such as Nebraska, Missouri, and Oregon, tend to have more stable funding and staffing to support conservation improvements.

The Pew Center on Global Climate Change has identified several ongoing state programs and demonstration projects specifically intended to promote carbon storage and emissions reduction in the U.S. agriculture sector. [59] For example, several states, including Oregon, Wisconsin, Vermont, and North Carolina, are promoting methane recovery and biofuels generation from livestock waste. A program in Iowa is providing support and funding to promote switchgrass as a biomass energy crop. In Maryland, state income tax credits are provided for the production and sale of electricity from certain biomass combustion. Georgia has a program that leases no-till equipment to farmers. In addition, several states, including Nebraska, Oklahoma, Wyoming, North Dakota, and Illinois, have formed advisory committees to investigate the potential for state carbon sequestration. In California, an accounting program is being developed to track possible future costs to mitigate GHG emissions in the U.S. agriculture sector.

State and Regional Climate Initiatives

Mandatory Programs

There are a number of state programs and initiatives geared toward climate change mitigation strategies across sectors including agriculture.[60] For example, the Center for Climate Strategies has assisted public officials in several states to develop climate action plans. Most of these plans incorporate strategies for emissions reduction goals in selected economic sectors, including the agriculture and forestry sectors. Plans for states such as Maryland, Michigan, and Florida include farm and forestry management activities ranging from forest and land use management to soil carbon management, tree planting, farmland conservation, expanded use of biomass feedstocks, methane capture and utilization, nutrient efficiency, and on-farm energy efficiency, among other practices. [61]

California is actively developing programs to support the state's enacted emission reductions legislation.[62] California's climate change statute requires state agencies to identify GHG emissions reduction strategies that can be pursued before most of the law takes effect in 2012. The state has identified several agriculture sector strategies that it plans to consider as early actions, including (1) adopting a manure digester protocol for calculating GHG mitigation; (2) establishing collaborative research on how to reduce GHG emissions from nitrogen land application; (3) replacing stationary diesel agricultural engines with electric motors; and (4) evaluating potential measures for enclosed dairy barns, modified feed management, and manure removal strategies to reduce methane emissions at dairies.[63] These early action strategies would be in addition to funding for the state's manure digester cost-share program and other agriculture projects, including carbon sequestration projects involving rice straw utilization, energy and water conservation, biofuels support, soil management, and other types of renewable energy and manure management programs for dairies.

Other regional climate initiatives include the Regional Greenhouse Gas Initiative (RGGI) and the Western Climate Initiative (WCI), among others. RGGI is a partnership of 10 northeastern and mid-Atlantic states that creates a cap-and-trade system aimed at limiting carbon dioxide emissions from power plants. Seven western states (and four Canadian provinces) have formed the WCI, which set an economy-wide GHG emissions target of 15% below 2005 levels by 2020.[64] Both RGGI and WCI include agricultural programs among their list of eligible offset and allowance project categories for trading emissions as part of their programs, along with other non-agricultural projects. Under RGGI, eligible agricultural and forestry project categories include sequestration of carbon due to afforestation, and avoided methane emissions from agricultural manure management operations.[65] Under WCI and California's climate statute, agriculture and forestry sector actions being considered for inclusion as offset and allowance projects cover forestry protocols, manure digester protocols, measures for enclosed dairy barns, modified feed management, manure removal strategies to reduce methane emissions at dairies, emission reductions from nitrogen land application, soil sequestration, and replacing stationary diesel agricultural engines with electric motors.[66]

Voluntary Carbon Market Programs

The voluntary carbon offset market allows businesses, interest groups, and individuals the opportunity to purchase carbon credits generated from projects that either prevent or reduce an amount of carbon entering the atmosphere, or that capture carbon from the atmosphere. Companies and individuals purchase carbon credits for varied reasons. For example, some may purchase credits to reduce their "carbon footprint," using credits to offset all or part of a GHG-emitting activity (e.g., air travel, corporate events, or personal automobile use); others may purchase credits to bank the reductions in anticipation of a mandatory GHG reduction program.[67] In the United States, the current offset framework operates on a voluntary basis since there is no federal requirement that GHG emissions be curtailed. Some states and/or regional GHG reduction initiatives may limit the use of carbon offsets.

Several states have programs that support the voluntary carbon offset exchange, often involving U.S. farmers and private landowners. Farmer participation in voluntary carbon credit trading programs has been growing rapidly. As of mid-2009, participation involved an estimated roughly 10,000 farmers across about 35 states covering more than more than 10 million acres.[68] One program, operated by the National Farmers Union (NFU), involves more than 4,000 producers in more than 30 states, with more than 5 million acres of farmland enrolled. Another program operated by the Iowa Farm Bureau involves 5,000 to 6,000 producers also in more than 30 states (mostly Iowa, Kansas, and Nebraska, but also Illinois, Ohio, Michigan, Wisconsin, Minnesota, South Dakota, Missouri, Indiana, and Kentucky), also with more than 5 million acres of farmland enrolled. The types of practices covered by this program include no-till crop management; conversion of cropland to grass; managed forests, grasslands, and rangelands; new tree plantings; anaerobic digesters and methane projects; wind, solar, or other renewable energy use; and forest restoration. Similar programs also have been initiated in Illinois (Illinois Conservation and Climate Initiative), Indiana (Environmental Credit Corporation), and the Northwest (Upper Columbia Resource Conservation and Development Council). Another, Terrapass, has among its projects two large-scale dairy farms that use anaerobic digesters and methane capture for energy production.[69]

These programs "aggregate" carbon credits across many farmers and landowners. These credits may later be sold on the Chicago Climate Exchange.[70] Farmer participation in such programs may help offset farm costs to install emissions controls and/or practices that sequester carbon by providing a means for them to earn and sell carbon credits.

Congressional Action

Energy and Climate Legislative Proposals

Congress is currently considering a range of energy and climate policy options. In general, the current climate proposals would not require GHG emission reductions in the agriculture and forestry sectors. However, if enacted, provisions in these bills could potentially raise farm input costs for fossil fuels, fertilizers, energy, and other production inputs. These higher costs could potentially be offset by possible farm revenue increases should farmers participate in carbon offset and renewable energy provisions that are part of this legislation. For example, within capand-trade proposals being debated in Congress are provisions that could provide tradeable allowances to certain agricultural industries, and provisions that could establish a carbon offset program for domestic farm- and land-based carbon storage activities. In addition, the renewable energy provisions contained in these bills could potentially expand the market for farm-based biofuels, biomass residues, and dedicated energy crops. These and related bills and issues are currently being debated in Congress. More detailed information on these bills is provided in other CRS Reports.

2008 Farm Bill Provisions

The omnibus 2008 farm bill (Food, Conservation, and Energy Act of 2008, P.L. 110-246) included a new ecosystem services market provision that expanded the scope of existing farm and forestry conservation programs in ways that could more broadly encompass certain aspects of these climate change initiatives. The 2008 farm bill's so-called environmental services market provision seeks to facilitate the participation of farmers and landowners in environmental services markets, focusing first on carbon storage .[71] This provision was also intended to help address some of the measurement and quantification issues surrounding agricultural and forestry carbon credits, as well as to expand existing voluntary conservation and other farm bill programs, providing incentives that could accelerate opportunities for agriculture and forestry to reduce emissions associated with climate change, adopt energy efficiency measures, and produce renewable energy feedstocks.

The 2008 farm bill provision seeks to "establish technical guidelines that outline science-based methods to measure the environmental services benefits from conservation and land management activities in order to facilitate the participation of farmers, ranchers, and forest landowners in emerging environmental services markets." The intended purpose of these technical guidelines is to develop (1) a procedure to measure environmental services benefits; (2) a protocol to report these benefits; and (3) a registry to collect, record, and maintain the benefits measured. The provision also requires that USDA provide guidelines for establishing a verification process as part of the protocol for reporting environmental services, but it allows USDA to consider the role of third parties in conducting independent verification. In carrying out this directive, USDA is directed to work in consultation with other federal and state government agencies, nongovernmental interests, and other interested persons as determined by USDA. However, the enacted bill did not specifically address funding for this provision. Nevertheless, the inclusion of this provision in the farm bill is expected to expand the scope of existing farm and forestry conservation programs in ways that will more broadly encompass certain aspects of the climate change debate. For more detailed background information, see CRS Report RL34042, *Provisions*

Supporting Ecosystem Services Markets in U.S. Farm Bill Legislation

In addition, this farm bill provision is invoked within the current energy and climate legislation. For example, H.R. 2454 would expand this provision to establish an independent advisory committee to provide advice on establishing and implementing a carbon offset program for domestic agricultural and forestry practices.[72]

In December 2008, USDA announced it would create a federal government-wide "Conservation and Land Management Environmental Services Board" to assist USDA with the "development of new technical guidelines and science-based methods to assess environmental service benefits which will in turn promote markets for ecosystem services including carbon trading to mitigate climate change."[73] A federally chartered public advisory committee will advise the board, and will include farmers, ranchers, forest landowners, and tribal representatives, as well as representatives from state natural resource and environmental agencies, agriculture departments, and conservation and environmental organizations. USDA's press release also announced that USDA was establishing a new Office of Ecosystem Services and Markets (OESM), which will provide administrative and technical assistance in developing the uniform guidelines and tools needed to create and expand markets for ecosystem services in the farming and forestry sectors.

Considerations for Congress

In March 2009, the House Agriculture Committee issued a climate change questionnaire, which was distributed to more than 400 organizations, to solicit input on proposals to reduce GHG emissions. The published survey responses are available on the committee's website and highlight some concerns, as well as the potential market opportunities issues for farmers and landowners.[74] These and other issues were discussed at a House Agriculture Committee hearing in June 2009, and also at subsequent Senate Agriculture Committee hearings in July and September 2009 as part of the committees' review of pending climate legislation.[75] Similar issues were raised at a 110th Congress subcommittee hearing of the Senate Agriculture Committee in May 2008.[76]

Although the current legislative proposals do not specifically include agricultural operations among "covered entities" under a mandatory emissions cap, some interest groups continue to question whether certain types of agricultural operations could eventually be brought in under some proposals. Still others continue to argue that U.S. agriculture will be affected by anticipated climate legislation in terms of generally increasing energy and production input costs that will negatively impact the farming sector.[77]

The extent to which the agricultural and forestry sectors will participate in an offset and allowance program continues to be actively debated in Congress. The inclusion of provisions that allow for agriculture and forestry offsets and allowances as part of a cap-and-trade scheme has been generally supported by a broad-based industry coalition. This coalition consists of agricultural groups representing commodity crops, livestock and dairy, the American Farm Bureau Federation, the National Farmers Union, the American Farmland Trust, and other agriculture support and utility companies.[78] Former Senators and Majority Leaders Bob Dole and Tom Daschle are also advocating on behalf of the Bipartisan Policy Center that farmers be fully integrated into any cap-and-trade program.[79] Most groups, including many within the environmental community, generally support the inclusion of carbon offset projects within a capand-trade scheme since this is likely to help contain overall costs of a carbon reduction program.

The inclusion of agriculture and forestry offsets with a carbon reduction program, however, has remained controversial since the Kyoto Protocol negotiations during the 1990s.[80] During those negotiations, there was marked disagreement among countries and interest groups, arguing either for or against the inclusion of offsets from the agriculture and forestry sectors.[81]

The EU's GHG emission program, the Emission Trading System (ETS), which was established in 2005, does not provide for agricultural or forestry projects and activities. Among the reasons are (1) pragmatic concerns regarding measurement and verification, given the sheer number of farmers and landowners, and (2) ideological concerns about granting too much flexibility in how emission reductions are met, which could undermine overall program goals. Among the areas of concern regarding biological sequestration offsets are those highlighted in two previous sections of this report, "Uncertainty Estimating Emissions" and "Uncertainty Estimating Carbon Sinks." In summary, primary areas of concern include

- *Permanence/Duration*, given that land uses can change over time (e.g., forest lands to urban development, other natural events such as fires or pests);
- *Measurement/Accounting,* given that biological sequestration measurement is difficult and estimates can vary, actual emission reduction or sequestration depends on site-specific factors (e.g., location, climate, soil type, crop/vegetation, tillage practices, farm management, etc.);
- *Effectiveness,* the success of the mitigation practice will depend on the type of practice, how well implemented and managed by the farmer or landowner, and the length of time the practice is undertaken;
- *Additionality/Double Counting,* given that some of the activities generating offsets would have occurred anyway under a pre-existing program or practice, and thus may not go beyond business as usual (BAU); and/or given that some reductions may be counted by another program (e.g., attributable to other environmental goals under various farm conservation programs) or toward more than one GHG reduction target; and
- *Leakage,* given that reductions in one place could result in additional emissions elsewhere.

A more detailed discussion of some of these issues is available in various reports by CRS,[82] the Government Accountability Office (GAO),[83] and other groups.

Following is a list of questions that may be raised as Congress continues to consider the role of the agriculture and forestry sectors as part of the broader climate change debate.

- *Emissions reductions.* Should carbon sequestration efforts be balanced by incentives to obtain additional emissions reductions in the agriculture sector through improved conservation and farm management practices, which could have a more immediate, direct, and lasting effect on overall GHG emissions? How might the existing regulatory framework for controlling air pollutants affect the climate change debate? What are the potential options for reducing GHG emissions at U.S. farming operations? How might cost concerns be addressed that limit broader adoption of manure management systems and also feed management strategies at U.S. livestock operations?
- *Carbon sequestration.* What are the upper limits of carbon capture and storage initiatives in the agriculture sector? For example, are such carbon sinks temporary or long-lasting, and what limits exist on their storage value? Do they rely appropriately on the willingness of landowners to adopt or continue to implement a particular conservation practice? Do they rely too heavily on the willingness of landowners to convert existing farmland to open space or prevent the conversion of existing farmland to non-farm uses? Are they cost-effective when compared to sinks in other sectors? How might concerns regarding uncertainty be addressed when measuring and estimating the amount of carbon sequestered in agricultural soils?
- *Carbon offset or credit markets.* What is the federal role in possibly expanding existing conservation programs in conjunction with efforts to create new market opportunities for farmers by developing a carbon credit trading system? How will USDA implement the new 2008 farm bill provision directing the Department to work with other agencies and organization to establish guidelines and standards for measuring agricultural and forestry environmental benefits, including carbon storage? What are the potential measurement, monitoring, enforcement, and administrative issues of implementing a carbon credit trading system involving the agriculture and forestry sectors? How would stored carbon be measured and verified; how much compensation would be available and for how long; what are required management practices; and which accounting methodologies should be used? Would such a system operate under a voluntary or a mandatory framework?
- *Farm bill Programs.* Are there opportunities to expand existing federal conservation and land management programs to achieve greater emissions reduction and carbon sequestration in the

agriculture sector? How might emissions reduction and carbon sequestration be integrated with the many other goals of conservation programs, such as improved soil quality and productivity, improved water and air quality, and wildlife habitat? Which programs or practices are the most beneficial and cost-effective? Are there ways to rank applications from farmers under existing programs to grant a higher weight to proposals to address climate change goals? Are there existing state programs that effectively address climate change and could be adopted at the federal level?

- *Bioenergy promotion.* How might ongoing or anticipated initiatives to promote U.S. bioenergy production, such as corn-based or cellulosic ethanol, affect the options for land management or conservation strategies that could increase carbon uptake on agricultural lands and in agricultural soils? Might broader climate change goals be affected by increased agricultural production in response to corn-based ethanol? For example, might previously retired land be brought back into corn production or might this result in more intensive corn production, including fewer crop rotations and planting area setbacks, which could raise emissions and reduce the amount of carbon sequestered? Are there other competing commercial crops that might be used as a feedstock for ethanol that could also affect emissions and carbon uptake potential?
- *Energy efficiency.* What are the opportunities for improved on-farm energy efficiency and conservation? How might these be integrated into the broader framework on climate change mitigation in the agriculture sector?
- *Safeguarding U.S. agricultural production.* Among the possible effects of global climate change on agricultural production are increased climate variability and increased incidence of global environmental hazards, such as drought and/or flooding, pests, weeds, and diseases, or location shifts in where agriculture is produced. Climate change in some locations increases the yields of some crops. Some U.S. production regions are likely to fare better than others. Are additional initiatives needed in the U.S. agriculture sector to prepare for the potentially effects of global climate change that might impact U.S. agricultural production and food security? Which regions and crops might be "winners" or "losers" and how can transitions be eased?

Appendix. Primer on Agriculture's Role in the Climate Change Debate

Question	Discussion
What are the types of GHG emissions associated with U.S. agriculture?	Official estimates of greenhouse gas (GHG) emissions for the U.S. agriculture sector are based on emissions of methane (CH_4), nitrous oxide (N_2O) associated with agricultural production, and carbon dioxide (CO_2) emissions from on-farm energy use. These estimates do not include other emissions associated with forestry activities, food processing or distribution, or biofuel production. See "Agricultural Emissions" in this report for more information.
What are the sources of GHG emissions from agriculture?	Agricultural sources of CH4 emissions are mostly associated with the natural digestive process of animals and with manure management on U.S. livestock operations. Sources of N_2O emissions are mostly associated with soil management and fertilizer use on U.S. croplands. *Figure 1* shows agricultural emissions by type and production category.
What is agriculture's share of annual national GHG emissions?	In the last five years, GHG emissions from U.S. agricultural activities have averaged nearly 514 MMTCO_2-Eq (million metric tons CO_2-equivalent units), accounting for about 7% of annual national GHG emissions (*Table 1*). Fossil fuel combustion is the leading source of national GHG emissions (about 80%), with the energy sector generating about 85% of annual emissions across all U.S. sectors.
How much carbon is sequestered in U.S. agricultural soils?	In the last five years, agricultural soils have sequestered, on average, about 44 MMTCO_2-Eq., or roughly 5% of annual emissions generated from agricultural activities. Compared to total national GHG emissions, the agriculture sector offsets well under 1% of emissions annually. These estimates do not include uptake from forested lands or open areas that account for a majority (about 95%) of total U.S. sequestration.

Is there uncertainty associated with estimates of carbon uptake for the agriculture sector?	Factors accounting for uncertainty in uptake estimates in U.S. soils include accounting methodology; type of practice, how well it is implemented, and the length of time undertaken; availability of federal/state cost-sharing or technical assistance; and other competing factors (including supply response for commercial crops and bioenergy crops). Actual GHG emissions may also vary according to many site-specific conditions (e.g., location, climate, soil type, crop type, tillage practices, crop rotations, farm management, etc.). See "*Uncertainty Estimating Carbon Sinks*" for more information.
What is the potential to reduce emissions and/or increase carbon uptake in the agriculture sector?	The potential for carbon uptake in the U.S. agriculture sector is much greater than current rates. USDA and EPA estimate net increases in carbon sequestration ranging from 590 to 990 MMTCO_2-Eq. per year (*Table 2*). This could offset total current national GHG emissions by as much as 8%-14%. Practices that may reduce emissions and/or sequester carbon on U.S. farmlands include land retirement, pastureland and crop conversion, and restoration; improved soil management and conservation tillage; and improved manure management and feeding strategies at livestock operations. See sections "*Potential for Additional Uptake" and "Potential for Additional Emission Reductions.*"
Are there existing programs and/or legislation that promote farming practices that may help address climate change?	Existing federal and state farm conservation programs promote the types of land management and conservation practices that can reduce GHG emissions and/or sequester carbon. Also, many existing voluntary programs in the current farm bill, as well as under existing state-level programs, provide cost-sharing and technical assistance to encourage farmers to implement such practices. These voluntary programs are generally designed to address site-specific improvements at an individual farming operation. See "*Federal Programs*" and other listed program information.

Source: Table prepared by the Congressional Research Service.

End Notes

[1] See CRS Report RL33849, *Climate Change: Science and Policy Implications.*

[2] See CRS Report RL34265, *Selected Issues Related to an Expansion of the Renewable Fuel Standard (RFS).*

[3] EPA *Inventory.*

[4] USDA, *U.S. Agriculture and Forestry Greenhouse Gas Inventory: 1990-2005*, TB1921, Figure 3-8, August 2008, at http://www.usda.gov/oce/global_change/ AFGGInventory 1990_ 2005.htm (hereafter "USDA *Inventory*").

[5] "Carbon-equivalents" equate an amount of a GHG to the amount of carbon that could have a similar impact on global temperature. EPA's data are in teragrams (million metric tons). Alternative ways to express emissions and offsets are in carbon equivalents (MMTCE) (assumes a multiplier of 0.272 to convert from EPA-reported equivalent CO_2-Eq. units).

[6] EPA *Inventory,* Table 2-14.

[7] Aside from the energy and agriculture/forestry sectors, by source, other leading contributors are wood biomass/ethanol use; nonenergy use of fuel; landfills; and substitution of ozone-depleting substances. By sector, leading sources are industrial processes and wastes. EPA *Inventory,* Tables ES-2 and ES-4.

[8] The principal gases associated with climate change from human activities are CO_2, CH_4, N_2O, and ozone-depleting substances and chlorinated and fluorinated gases, such as hydrofluorocarbons, perfluorocarbons, and sulfur hexafluoride. See CRS Report RL33849, *Climate Change: Science and Policy Implications.*

[9] Methane's ability to trap heat in the atmosphere is 21 times that of CO2; nitrous oxide is 310 times that of CO_2 (measured over a 100-year period). Intergovernmental Panel on Climate Change (IPCC), *Climate Change 2007,* Technical Summary of the Working Group I Report, Table TS-2, at http://ipcc-wg1.ucar. edu/wg1/Report/ AR4WG1_Print_TS.pdf.

[10] USDA *Inventory*, Figure 3-6. Nitrogen-fixing crops refer to beans, legumes, alfalfa, and non-alfalfa forage crops.

[11] EPA *Inventory,* Table ES-2. Based on five-year average of available data. Other major CH4 sources were landfills, natural gas systems, and coal mining. Mobile combustion was the second largest source of N_2O.

[12] EPA *Inventory,* Table 2-14.

[13] Also see CRS Report R40874, *Nitrous Oxide from Agricultural Sources: Potential Role in Greenhouse Gas Emission Reduction and Ozone Recovery.*

[14] Refers to livestock (cattle, sheep, goats, and buffalo) that have a four-chambered stomach. In the rumen chamber, bacteria breaks down food and degrades methane as a byproduct.

[15] R. A. Leng, "Quantitative Ruminant Nutrition—A Green Science," *Australian Journal of Agricultural Research*, 44: 363-380. Feed efficiency based on both fermentive digestion in the rumen and conversion of feed to output (e.,g, milk, meat) as nutrients are absorbed.

[16] Although carbon is released as well, it is predominantly absorbed again within a year as part of the cropping cycle, and so is assumed to be net zero emissions unless some goes into long-term soil carbon content.

[17] Previously estimates for the agriculture soil management category were lower. Current EPA estimates reflect methodological and input data changes.

[18] EPA *Inventory,* Table 2-14.

[19] EPA *Inventory,* Table 6-2. NOX and CO influence the levels of tropospheric ozone, which is both a local pollutant and a GHG (called "indirect" greenhouse gases). Their contributions cannot be measured by emissions.

[20] See CRS Report RL32948, *Air Quality Issues and Animal Agriculture: A Primer*. Particulate emissions may also contribute to climate change, but their influence is predominantly local, short-term and poorly quantified.

[21] The IPCC was established to assess scientific, technical and socioeconomic information related to climate change, its potential impacts and options for adaptation and mitigation. IPCC's methodology to estimate emissions and sinks are consistent with those used by other governments and with established guidelines under the United Nations Framework Convention on Climate Change.

[22.] D. C. Reicosky, "Environmental Benefits of Soil Carbon Sequestration," USDA, http://www.dep.state.pa.us/dep/DEPUTATE/Watermgt/wsm/WSM_TAO/InnovTechForum/InnovTechForum-IIE-Reicosky.pdf.

[23] USDA, "Conservation Tillage Firmly Planted in U.S. Agriculture," *Agricultural Outlook*, March 2001; USDA, "To Plow or Not to Plow? Balancing Slug Populations With Environmental Concerns and Soil Health," *Agricultural Research,* October 2004; Conservation Technology Information Center (CTIC), "Conservation Tillage Facts," at http://www.conservationinformation.org/?action=learningcenter_core4_convotill.

[24] An enclosed tank that promotes decomposition using anaerobic conditions and naturally occurring bacteria, while producing biogas as a byproduct that can be used as energy.

[25] See CRS Report RL32948, *Air Quality Issues and Animal Agriculture: A Primer.*

[26] See CRS Report R40667, *Anaerobic Digestion: Greenhouse Gas Emission Reduction and Energy Generation*; and R. Pillars, "Farm-based Anaerobic Digesters," Michigan State University Extension, http://web2.msue.msu.edu/ manure/FinalAnearobicDigestion Factsheet.pdf.

[27] EPA, *Development Document for the Final Revisions to the NPDES Regulation and the Effluent Guidelines for Concentrated Animal Feeding Operations*, January 2003.

[28] C. Henry and R. Koelsch, "What Is an Anaerobic Digester?" University of Nebraska, Lincoln, at http://files.harc.edu/ Sites/GulfcoastCHP/Publications/WhatIsAnaerobicDigestion.pdf; and Pennsylvania State University, "Biogas and Anaerobic Digestion," at http://www.biogas.psu.edu/. For optimum operation, anaerobic digesters must be kept at a constant, elevated temperature, and any rapid changes in temperature could disrupt bacterial activity.

[29] Previously, mostly under Section 9006 and Section 6013 of the 2002 farm bill (P.L. 107-171), but also under other farm bill cost-share programs. CRS communication with USDA staff.

[30] As of 2005. EPA, *AgStar Digest*, Winter 2006, at http://www.epa.gov/agstar/.

[31] R. A. Leng, "Quantitative Ruminant Nutrition—A Green Science," *Australian Journal of Agricultural Research*, 44: 363-380; H. Steinfeld, C. de Haan, and H. Blackburn, *Livestock-Environment Interactions, Issues and Options*, chapter 3 (study commissioned by the Commission of the European Communities, United Nations, and World Bank), at http://www.virtualcentre.org/es/dec/toolbox/FAO/Summary/index.htm.

[32] U.S. Geological Survey (USGS), website information on carbon sequestration in soils.

[33] Table 2 of CRS Report RS22964, *Measuring and Monitoring Carbon in the Agricultural and Forestry Sectors*, summarizes information in EPA, *Greenhouse Gas Mitigation Potential in U.S. Forestry and Agriculture*, Nov. 2005, http://www.epa.gov/sequestration/ greenhouse_gas.html, and USDA, *Economics of Sequestering Carbon in the U.S. Agricultural Sector*, Apr. 2004, http://www.ers.usda. gov/publications/tb1909/. The estimates show the potential for carbon storage (tonnage) by type of farming and forestry activity.

[32] U.S. Geological Survey (USGS), website information on carbon sequestration in soils.

[33] Table 2 of CRS Report RS22964, *Measuring and Monitoring Carbon in the Agricultural and Forestry Sectors*, summarizes information in EPA, *Greenhouse Gas Mitigation Potential in U.S. Forestry and Agriculture*, Nov. 2005, http://www.epa.gov/sequestration/ greenhouse _gas.html, and USDA, *Economics of Sequestering Carbon in the U.S. Agricultural Sector*, Apr. 2004, http://www.ers.usda.gov/publications/tb1909/. The estimates show the potential for carbon storage (tonnage) by type of farming and forestry activity.

[34] Most current carbon sequestration is within the forestry sector (see *Table 1*).

[35] EPA *Inventory,* Table 2-12. Based on estimates for CO_2 flux in agricultural soil carbon stocks.

[36] EPA *Inventory,* Table 2-12 and Table 7-1. Based on estimates for the following categories: forestland remaining forestland; and growth in urban trees. Other uptake not included in the estimates is from landfilled yard trimmings.

[37] Congressional Budget Office (CBO), *The Potential for Carbon Sequestration in the United States*, Sept. 2007, at http://www.cbo.gov/ftpdocs/86xx/doc8624/09-12-Carbon Sequestration.pdf. Biological sequestration refers to the use of land to enhance its ability to uptake carbon from atmosphere through plants and soils. Direct sequestration refers to capturing carbon at its source and storing it before its release to the atmosphere. Examples include capture and storage in geologic formations, such as oil fields, natural gas fields, coal seams, and deep saline formations. See CRS Report RL33801, *Carbon Capture and Sequestration (CCS).*

[38] See, for example, T. A. Butt and B. A. McCarl, "Implications of Carbon Sequestration for Landowners," *2005 Journal of the American Society of Farm Managers and Rural Appraisers*; Government Accountability Office (GAO), *Conservation Reserve Program: Cost-Effectiveness Is Uncertain*, March 1993; H. Feng, J. Zhao, and C. Kling, "Carbon: The Next Big Cash Crop," *Choices*, 2nd quarter 2001; and H. Feng, C. Kling, and P. Glassman, "Carbon Sequestration, Co-Benefits, and Conservation Programs," *Choices*, Fall 2004.

[39] W. Hohenstein, "USDA Activities to Address Greenhouse Gases and Carbon Sequestration," presentation to Senate Energy Committee staff, February 15, 2007.

[40] See, for example, statements by representatives of the American Farm Bureau Federation and the National Farmers Union to House Agriculture Committee staff, May 18, 2009.

[41] Net reduction below baseline at a range of carbon prices from about $3 to $30/MT CO_2-Eq., annualized assuming a 15-year program.

[42] Reported by EPA, *Greenhouse Gas Mitigation Potential in U.S. Forestry and Agriculture*, Nov. 2005, Tables 4-10 (15-year), http://www.epa.gov/sequestration/greenhouse_gas.html. The resultant estimates may overlap between the afforestation and forest management categories.

[43] CBO, *The Potential for Carbon Sequestration in the United States*, Sept. 2007, http://www.cbo.gov/ftpdocs/86xx/doc8624/09-12-CarbonSequestration.pdf. See CBO report Figures 1 and 2.

[44] See EPA memorandum, "Updated Forestry and Agriculture Marginal Abatement Cost Curves," March 31, 2009.

[45] EPA, "Waxman-Markey Discussion Draft Preliminary Analysis: EPA Preliminary Analysis of the American Clean Energy and Security Act of 2009," Appendix, http://epa.gov/climatechange/economics/pdfs/WM-Appendix.pdf. See slide 25-27 for agriculture and forestry modeling results. Other information on EPA's analysis is at http://epa.gov/ climatechange/economics/pdfs/WM-Analysis.pdf.

[46] USDA *Inventory,* Figure 3-8.

[47] Bongen, A.," Using Agricultural Land for Carbon Sequestration," Purdue University, at http://www.agry.purdue.edu/ soils/Csequest.PDF. 1999 data for carbon storage in Indiana.

48 USDA *Inventory*; and "Depositing Carbon in the Bank: The Soil Bank, That Is," *Agricultural Research,* Feb. 2001.

49 For example, see Iowa Farm Bureau's carbon credit project at http://www. Iowafarmbureau. com, and Environmental Credit Corporation at http://www.envcc.com.

50 For more information, see CRS Report RL31432, *Carbon Sequestration in Forests.*

51 USDA, "Reducing Agricultural Greenhouse Gas Emissions Through Voluntary Action," Statement by Bruce Knight of USDA's Natural Resources Conservation Service at the United Nations Framework Convention on Climate Change, December 2004, at http://www.nrcs.usda.gov/news/speeches04/climatechange.html.

52 The following program rules include a section recognizing the credits generated by programs and asserting no direct or indirect claim on these credits: EQIP (§1466.36, 74 *Federal Register* 2317), WRP (§1467.20, 74 *Federal Register* 2336), AMA (§1465.36, 73 *Federal Register* 70256), GRP (§1415.10, 74 *Federal Register* 3875), FPP (§1491.21, 74 *Federal Register* 2822), WHIP (§636.21, 74 *Federal Register* 2800), CRP (§1410.63(6), 68 *Federal Register* 24845), and HFRP (§625.8, 74 *Federal Register* 1967). Also see CRS Report R40692, *Agricultural Conservation Issues in the 111th Congress.*

53 A previous program in Title VI (Rural Development) that was not reauthorized in the 2008 farm bill was a provision (Section 6013) authorizing rural development business and industry program to make loans and loan guarantees for renewable energy systems, including wind energy systems and anaerobic digesters.

54 Previously referred to as Section 9006 (Renewable Energy Systems and Energy Efficiency Improvements) in the 2002 farm bill.

55 CRS communication with USDA staff, February 8, 2007. Limited information indicates that USDA funded eight projects totaling more than $60 million under the previous Section 6013 and provided another $20 million in funding assistance under Section 9006 for anaerobic digesters (FY2002-FY2005).

56 See CRS Report RL34130, *Renewable Energy Programs in the 2008 Farm Bill*; CRS Report RL32712, *Agriculture-Based Renewable Energy Production*; and CRS Report R40110, *Biofuels Incentives: A Summary of Federal Programs.*

57 A previous program that was not reauthorized in the 2008 farm bill was the Forest Service's Forest Land Enhancement Program (FLEP). FLEP provided funding for agriculture and silvopasture practices with rotational grazing and improved forage. Primary efforts under the program included afforestation and reforestation, improved forest stand, constructing windbreaks, and riparian forest buffers. For information on USDA forestry programs, see CRS Report RL33917, *Forestry in the 2008 Farm Bill.*

58 State and Local Government directory, http://www.statelocalgov.net/index.cfm; State Law Clearinghouse compilations by the National Agriculture Law Center, http://www. nationalaglawcenter.org/reporter/clearinghouse/.

59 Pew Center, *Learning from State Action on Climate Change*, Oct. 2006, http:// www. pewclimate.org/policy_center/policy_reports_and_analysis/state.

60 Also see CRS Report RL33812, *Climate Change: Action by States to Address Greenhouse Gas Emissions.*

61 See individual state action plans, "What's Happening: U.S. Climate Policy Action," http://climatestrategies.us/.

62 California's Global Warming Solutions Act of 2006 (AB 32), which was enacted in September 2006, codified the state's goal of requiring California's GHG emissions be reduced to 1990 levels by 2020.

[63] California EPA, "Expanded List of Early Action Measures to Reduce Greenhouse Gas Emissions in California Recommended for Board Consideration," October 2007, http://www.arb.ca.gov/cc/ccea/meetings/ea_final_report.pdf.

[64] For more detailed information, see CRS Report RL33812, *Climate Change: Action by States to Address Greenhouse Gas Emissions.*

[65] Non-agricultural project categories include landfill methane capture and destruction, reduction in emissions of sulfur hexafluoride (SF6), and reduction/avoidance of CO_2 emissions from natural gas, oil, or propane end-use combustion due to end-use energy efficiency in the building sector. RGGI, "Overview of RGGI CO_2 Budget Trading Program," Oct. 2007, at http://www.rggi.org/docs/program_summary_10_07.pdf.

[66] Non-agriculture actions being considered include SF6 reductions in the non-electric sector, energy efficiency and other changes at cement facilities, changes in production inputs for some consumer products and in some manufacturing sectors, changes in transportation and shipping facilities, and waste management (landfill gas and wastewater management). WGI, http://www.westernclimateinitiative.org/; and California EPA, "Expanded List of Early Action Measures to Reduce Greenhouse Gas Emissions in California Recommended for Board Consideration," October 2007, http://www.arb.ca.gov/cc/ccea /meetings/ea_ final_ report.pdf.

[67] For additional general information on voluntary carbon markets, see CRS Report RL34241, *Voluntary Carbon Offsets: Overview and Assessment.* For trading purposes, one carbon credit is considered equivalent to one metric ton of carbon dioxide emission reduced.

[68] Statements by the National Farmers Union and the Iowa Farm Bureau/AgraGate to House Agriculture Committee staff, May 18, 2009.

[69] For more information, see North Dakota Farmers Union at http://www.ndfu.org, Illinois Conservation and Climate Initiative at http://www.illinoisclimate.org, Environmental Credit Corporation at http://www.envcc.com; and Terrapass at http://www.terrapass.com/projects.

[70] The Exchange is a voluntary, self-regulated, rules-based exchange. Its emission offset program constitutes a small part of its overall program, which includes methane destruction, carbon sequestration, and renewable energy. See http://www.chicagoclimatex.com/.

[71] P.L. 110-246, Section 2709, included new language amending Section 1245(f) of the Food Security Act of 1985. Ecosystem services refers to the environmental goods and services and other benefits that the society obtains from the environment and ecosystems, both natural and managed. Examples include water filtration, flood control, provision of habitat, carbon storage, and many others. For more information, see CRS Report RL34042, *Provisions Supporting Ecosystem Services Markets in U.S. Farm Bill Legislation.*

[72] H.R. 2454, Sec. 531.

[73] USDA, "USDA Announces New Office of Ecosystem Services and Markets," Release No. 0307.08, Dec. 18, 2008.

[74] House Agriculture Committee's publications page is at http://agriculture.house.gov/inside/ publications.html.

[75] House Committee on Agriculture hearing, "To review pending climate legislation," June 11, 2009, http://agriculture.house.gov/hearings/statements.html; and Senate Committee on Agriculture, Nutrition and Forestry, "The role of agriculture and forestry in global warming legislation," July 22, 2009, and "Global Warming Legislation: Carbon Markets and Producer Groups," September 9, 2009, http://agriculture.senate.gov/.

[76] Subcommittee on Rural Revitalization, Conservation, Forestry and Credit hearing, May 21, 2008, "Creating Jobs with Climate Solutions: How agriculture and forestry can help lower costs in a low-carbon economy," at http://agriculture.senate.gov/.

[77] See, for example, statements by the American Farm Bureau at the 2009 USDA Outlook Forum, February 19, 2009; statements by various agriculture groups to House Agriculture Committee staff, May 18, 2009; and a study conducted for the Fertilizer Institute, at http://www.tfi.org/issues/climate/doanestudy.pdf.

[78] National Association of Wheat Growers, "Ag, Utility Groups Write on Stabenow Amendment," June 13, 2008, at http://www.wheatworld.org/html/news.cfm?ID=1423.

[79] Senators Bob Dole and Tom Daschle, *The Role of Agriculture in Reducing Greenhouse Gas Emissions: Recommendations for a National Cap-and-Trade Program*, April 2008, at http://www. bipartisanpolicy.org/ht/display/ArticleDetails/i/6086.

[80] See, for example, E. Boyd, E. Corbera, B. Kjellén, M. Guitiérrez, and M. Estrada, "The Politics of 'Sinks' and the CDM: A Process Tracing of the UNFCCC Negotiations (pre-Kyoto to COP-9)," Feb. 2007, draft submitted for International Environmental Agreements; also see two articles in *Nature*, no. 6812, Nov. 2000, "Deadlock in the Hague, but Hope Remains for Spring Climate Deal," and "Critical Politics of Carbon Sinks."

[81] Referred to as "land use, land use change, forestry," or abbreviated as LULUCF.

[82] CRS Report RS22964, *Measuring and Monitoring Carbon in the Agricultural and Forestry Sectors*, and CRS Report RL34436, *The Role of Offsets in a Greenhouse Gas Emissions Cap-and-Trade Program: Potential Benefits and Concerns.*

[83] GAO, *Carbon Offsets: U.S. Voluntary Market Is Growing, but Quality Assurance Poses Challenges for Market Participants*, GAO-08-1048, Aug. 29, 2008; GAO, *Climate Change: Observations on the Potential Role of Carbon Offsets in Climate Change Legislation*, GAO-09-456T, March 5, 2009; and GAO, *International Climate Change Programs: Lessons Learned from the European Union's Emissions Trading Scheme and the Kyoto Protocol's Clean Development Mechanism*, GAO-09-151 November 18, 2008, http://www.gao.gov/ products/ GAO-09-151.

In: Environmental Considerations in Agriculture ISBN: 978-1-61470-077-7
Editors: E.A. Sargsyan, A.J. Ochion 2011 Nova Science Publishers, Inc.

Chapter 4

NITROUS OXIDE FROM AGRICULTURAL SOURCES: POTENTIAL ROLE IN GREENHOUSE GAS EMISSION REDUCTION AND OZONE RECOVERY*

Kelsi Bracmort

SUMMARY

Gases other than carbon dioxide accounted for nearly 15% of total U.S. greenhouse gas emissions in 2008, yet there has been minimal discussion of these other greenhouse gases in climate and energy legislative initiatives. Reducing emissions from non-carbon dioxide greenhouse gases, such as nitrous oxide (N_2O), could deliver short-term climate change mitigation results as part of a comprehensive policy approach to combat climate change.

Nitrous oxide is 310 times more potent than carbon dioxide in its ability to affect the climate; and moreover, results of a recent scientific study indicate that nitrous oxide is currently the leading ozone-depleting substance being emitted. Thus, legislation to restrict nitrous oxide emissions could contribute to both climate change protection and ozone recovery.

* This is an edited, reformatted and augmented version of a Congressional Research Service 7-5700, R40874 publication, dated January 4, 2011.

The primary human source of nitrous oxide is agricultural soil management, which accounted for two-thirds of the N_2O emissions reported in 2008 (approximately 216 million metric tons CO_2 equivalent). One proposed strategy to lower N_2O emissions is more efficient application of synthetic fertilizers. However, further analysis is needed to determine the economic feasibility of this approach as well as techniques to measure and monitor the adoption rate and impact of N_2O emission reduction practices for agricultural soil management.

As the 112th Congress considers legislation that would limit greenhouse gas emissions, among the issues being discussed is how to address emissions of non-CO_2 greenhouse gases. Whether such emissions should be subject to direct regulation, what role EPA should play using its existing Clean Air Act authority, and what role USDA should play in any N_2O reduction scheme are among the issues being discussed. How these issues are resolved will have important implications for agriculture, which has taken a keen interest in climate change legislation.

INTRODUCTION

Policymakers are dedicating considerable attention to greenhouse gas emission reduction, primarily discussing options for carbon dioxide (CO_2) emission reduction.[1] Less frequently addressed in proposed legislation is emission reduction for non-CO_2 greenhouse gases, such as nitrous oxide (N_2O). However, N_2O reduction efforts have the potential to mitigate climate change. Moreover, N_2O emission sources may be regulated under the existing Clean Air Act as a class I or class II ozone-depleting substance at the discretion of the Environmental Protection Agency (EPA) Administrator. No new legislation needs to be passed to regulate N_2O for climate protection and ozone recovery.

The five non-CO_2 greenhouse gases regularly monitored but not entirely regulated by EPA (methane, nitrous oxide, hydroflourocarbons, perflourocarbons, and sulfur hexaflouride) accounted for nearly 15% of U.S. greenhouse gas (GHG) emissions in 2008, as measured by total tons of CO_2 equivalent.[2] Nitrous oxide—the third-most abundant greenhouse gas—was responsible for roughly 4% of total U.S. GHG emissions in 2008 by weight. Although they comprise a smaller portion of GHG emissions, non-CO_2 greenhouse gases, including N_2O, are more potent than CO_2. The gases identified above are 21 to 23,900 times more effective than an equivalent weight of CO_2 at trapping heat in the atmosphere, with N_2O being 310 times more potent by weight.[3]

In addition to being one cause of greenhouse gas emission growth, N_2O is an ozone-depleting substance (ODS).[4] Indeed, scientific analysis suggests that N_2O is now the leading ODS being emitted, as emissions of other substances have been reduced significantly owing to regulations enacted in the late 1980s, in the Montreal Protocol on Substances that Deplete the Ozone Layer.[5] N_2O emission reduction could thus play a compelling role in recovery of the ozone layer as well as in greenhouse gas emission reduction.

The agriculture sector is the primary anthropogenic source of nitrous oxide.[6] The bulk of U.S. N_2O emissions stem from fertilizing agricultural soils for crop production. Strategies or technologies designated for N_2O emission reduction are limited.[7] This is partly due to the dispersed nature of N_2O emission sources.

In the agriculture sector, the majority of N_2O is released as a consequence of specific nitrogen cycle processes (nitrification and denitrification) when large amounts of synthetic nitrogen fertilizers are used for crop production. More efficient application of synthetic fertilizers (e.g., precision agriculture, nitrogen inhibitors, nitrogen sensors, controlled-release fertilizer products) is one way to reduce excess amounts of nitrogen available for bacterial processing and eventual release to the atmosphere as N_2O. High costs and difficulty in measuring these products' efficacy, among other deterrents, have hampered widespread adoption of practices to reduce N_2O emissions.

This report focuses on the contributions of N_2O to alteration in the Earth's climate and ozone depletion. Policy options for N_2O emission reduction, sources of N_2O, and federal support to lower N_2O emissions are discussed.

NITROUS OXIDE: A PRIMER

Nitrous oxide (N_2O), familiar to some as "laughing gas," contributes to climate change and ozone depletion. Once released, N_2O lingers in the atmosphere for decades (its atmospheric lifetime is approximately 114 years) and is 310 times more effective at trapping heat in the atmosphere over a 100-year time frame than carbon dioxide (CO_2).[8] N_2O emission quantity estimates have remained fairly constant over the last few years, hovering around 325 million metric tons carbon dioxide equivalent (CO_2e). See *Table 1.*

Table 1. U.S. Greenhouse Gas Emissions

(million metric tons CO_2e)

Gas / Source	2006	2007	2008	Avg. Contribution[a]
Carbon dioxide (CO_2)	6,017.2	6,120.2	5,921.2	85%
Methane (CH_4)	568.2	569.2	567.6	8%
Nitrous oxide (N_2O)	329.5	327.7	318.2	4.6%
Hydroflourocarbons (HFCs)	121.8	127.4	126.9	1.7%
Perflouruocarbons (PFCs)	6.0	7.5	6.7	<1%
Sulfur hexaflouride (SF6)	17.0	16.1	16.1	<1%
Total	7,059.7	7,168.1	6,956.7	

Source: Environmental Protection Agency, 2010 U.S. Greenhouse Gas Inventory Report.

[a]. Average contribution to total U.S. greenhouse gas inventory based on data provided for 2006 to 2008.

Sources of N_2O Emissions

Nitrous oxide is emitted from anthropogenic (manmade) and natural sources. Oceans and natural vegetation are the major natural sources of N_2O. Agricultural soil management (e.g., fertilization, application of manure to soils, drainage and cultivation of organic soils) is responsible for two-thirds of anthropogenic U.S. N_2O emissions.[9] In 2008, N_2O emissions from agricultural soil management totaled more than 200 million metric tons of CO_2e.[10] Other anthropogenic sources of N_2O are combustion by mobile sources (cars, trucks, etc.), nitric acid production, and manure management.[11]

Figure 1 depicts the origination and passage of nitrogen (N) that leads to N_2O emissions from agricultural soil management. The amount of N_2O emitted from cropland soils largely depends on the amount of nitrogen applied to a crop, weather, and soil conditions. Corn and soybean crops emit the largest amounts of N_2O, respectively, due to vast planting areas, plentiful synthetic nitrogen fertilizer applications, and, in the case of soybeans, high nitrogen fixation rates (*Figure 2*).[12]

The Nitrogen Cycle

Comprehension of the nitrogen cycle (*Figure 3*) is beneficial when crafting policy to reduce N_2O emissions from anthropogenic sources. Nitrogen, an essential element required by organisms to grow, is found throughout the atmosphere in various forms. The nitrogen cycle portrays the routes in which nitrogen moves through the soil and atmosphere in both organic and inorganic form. Certain processes within the nitrogen cycle convert the nitrogen into a form that can be taken up by plants. Four of the major processes are:

- nitrogen fixation—conversion of nitrogen gas (N2) to a plant-available form;
- nitrogen mineralization—conversion of organic nitrogen to ammonia (NH3);
- nitrification—conversion of ammonia (NH3) to nitrate (NO_3-) via oxidation (that is, by being combined with oxygen); and
- denitrification—conversion of nitrates back to nitrogen gas.

Nitrous oxide is a byproduct of nitrification and denitrification. Both processes occur naturally. Excess application of nitrogen fertilizer can lead to increased nitrification, which can cause nitrate to leach into groundwater or surface runoff (in turn, this causes eutrophication, which can damage aquatic environments).

Opportunities and Challenges for Nitrous Oxide Emission Reduction

N_2O emission mitigation options are available for agricultural soil management and nitric acid production. Nitric acid is a chemical compound used to make synthetic fertilizers.

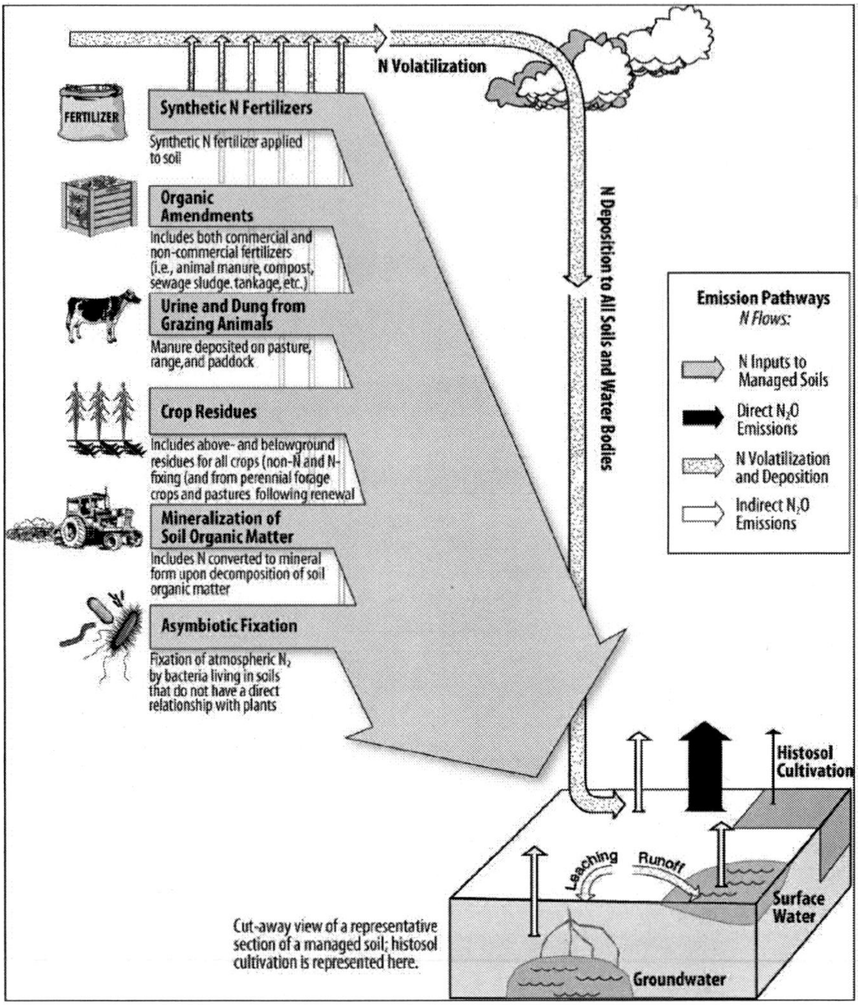

Source: U.S. Environmental Protection Agency, 2009 U.S. Greenhouse Gas Inventory Report, EPA 430-R-09-004, Chapter 6, April 2009. Adapted by CRS.

Figure 1. Sources and Pathways of Nitrogen (N) Resulting in N_2O Emissions from Agricultural Soil Management.

N_2O abatement options for nitric acid production include a high-temperature catalytic reduction method, a low-temperature catalytic reduction method, and nonselective catalytic reduction.[13]

The estimated reduction efficiencies (the percentage reduction achieved with adoption of a mitigation option) are 90%, 95%, and 85%, respectively.

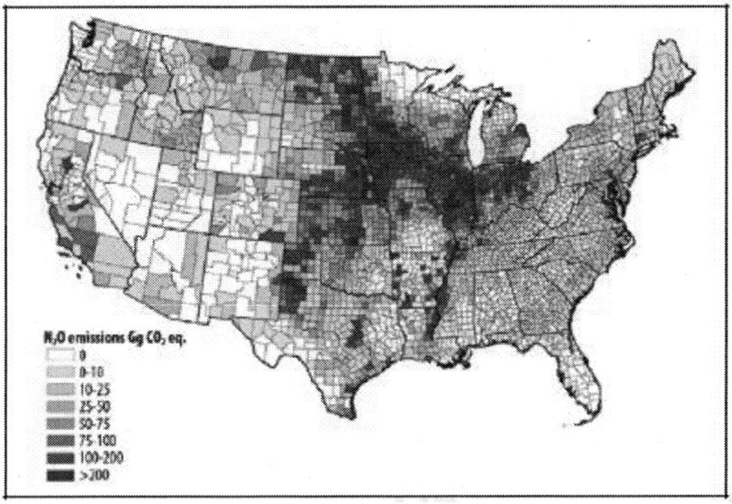

Source: USDA, U.S. Agriculture and Forestry Greenhouse Gas Inventory: 1990-2005. Adapted by CRS.

Notes: 1 Gigagram (Gg) is equivalent to 1,000 metric tons. Major crops are defined as corn, soybean, wheat, hay, sorghum, and cotton.

Figure 2. County-Level N_2O Emissions from Major Cropped Soils in 2005.

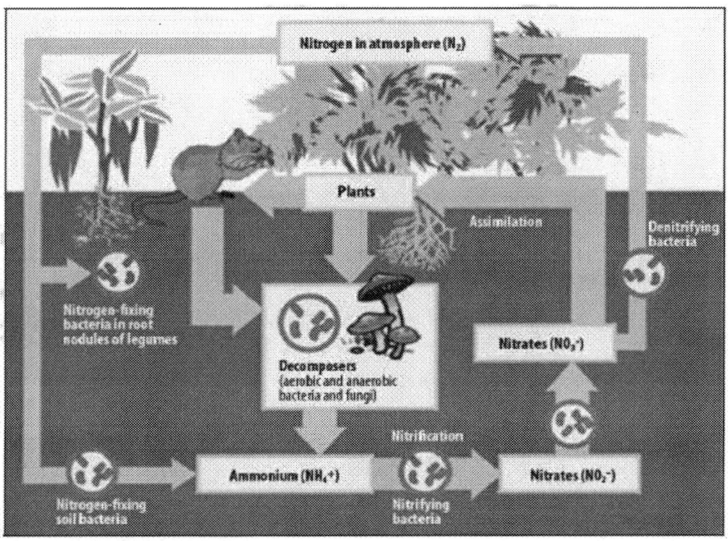

Source: EPA. Adapted by CRS.

Figure 3. The Nitrogen Cycle.

Agricultural soil management mitigation options recommended by researchers and technology transfer specialists to discourage excess application of nitrogen fertilizers and soil disturbance (*Table 2*) are not generally being practiced.

Fertilizer and soil best-management practices aim to provide the crop with the nutrient and soil conditions necessary for crop production, and prevent nutrient and soil loss from the crop field (e.g., erosion, leaching).[14] Some may consider less money spent towards fertilizer use an economic incentive for agricultural producers.[15] Others may want to ensure that crop yields meet expected feed, fiber, and fuel mandates (e.g., for corn ethanol), which may be difficult to attain with less fertilizer use.[16] Monitoring reduced nitrogen fertilization applications on a large scale for greenhouse gas emission reduction purposes may be difficult; it is not clear how such a program could be managed at a national level.[17]

Table 2. Select N_2O Mitigation Alternatives for Agricultural Soil Management

Mitigation Alternative	Description
Split fertilization	Application of same amount of nitrogen fertilizer as in baseline but divided into three smaller increments during crop uptake period to better match nitrogen application with crop demand and reduce nitrogen availability for leaching, nitrification, denitrification, and volatilization.
Simple fertilization reduction (10%, 20%, or 30%)	Reduction of nitrogen-based fertilizer from one-time baseline application of 10%, 20%, or 30%.
Nitrification inhibitor	Reduces conversion of ammonium to NO3-, which slows the immediate availability of nitrate (nitrate is water soluble). The inhibition of nitrification reduces nitrogen loss and increases overall plant uptake.
No-till	Conversion from conventional tillage to no-till, where soils are disturbed less and more crop residue is retained.

Source: EPA, Global Mitigation of Non-CO_2 Greenhouse Gases, http://www.epa.gov/climatechange/economics downloads/GM_SectionV_Agriculture.pdf.

Enforcement options could include voluntary verification, third-party verifiers, or government intervention. Reporting N_2O emissions from agricultural soil management was not included in the Final Mandatory Reporting of Greenhouse Gases Rule issued by EPA on September 22, 2009.[18]

EPA's reasoning behind this decision was that no low-cost or simple direct N_2O measurement methods exist. Additionally, EPA released a proposed rule requiring new or modified facilities that could trigger Prevention of Significant Deterioration (PSD) permitting requirements to apply for a revision to their operating permits to incorporate the best available control technologies and energy efficiency measures to minimize GHG emissions.[19]

FEDERAL SUPPORT FOR NITROUS OXIDE EMISSION REDUCTION

USDA provides some financial and technical assistance for nutrient management through its conservation programs.[20] Moreover, USDA's Agricultural Research Service (ARS) is studying the relationship between agricultural management practices and nitrous oxide emissions.[21]

In addition to the agriculture sector, work is being done in the transportation sector to reduce N_2O emissions. Mobile combustion was responsible for roughly 8% of N_2O emissions reported in 2008.[22] One N_2O emission reduction effort, proposed by EPA and the Department of Transportation, is a per-vehicle N_2O emission standard of 0.010 grams per mile effective in model year 2012 for all light-duty cars and trucks as part of a wider effort to reduce greenhouse gas emissions and improve fuel economy in tandem.[23] EPA has allocated financial resources to quantify N_2O emissions for the greenhouse gas inventory (e.g., DAYCENT model).[24]

POLICY OPTIONS FOR NITROUS OXIDE EMISSION REDUCTION

Congress has begun to investigate the reduction of non-CO_2 greenhouse gas emissions, including N_2O emissions, as one strategy to mitigate climate change. Some contend that N_2O emissions reduction could serve as a short-term response in the larger, long-term scheme of mitigation and adaptation efforts.[25] It may be viewed as a short-term response because N_2O emissions make up a small amount of the GHG inventory compared to CO_2 emissions. Any substantial approach to mitigate climate change is likely at some point to have to address sources that emit CO_2.

Congress could approach N_2O emissions reduction as part of a comprehensive GHG emission strategy offering economically attractive abatement alternatives to discourage actions leading to climate change. For example, a cap or fee on N_2O emissions could spur innovative methods for agricultural producers to limit excess synthetic fertilizer application. Congress could also examine the tools necessary to identify N_2O emission abatement options, assess their cost, and determine their economic impact for full incorporation into climate change legislation.

Besides greenhouse gas emission reduction, reducing N_2O emissions could lead to ozone recovery. Congress could explore the co-benefits that may arise from restricting N_2O emissions for climate change purposes. N_2O is not regulated as an ODS under the Clean Air Act, Title VI, Stratospheric Ozone Protection (as guided by the Montreal Protocol). As emissions of other ODSs (e.g., chlorofluorocarbon-11, halon-1211) have declined due to regulation, N_2O has emerged as the dominant ODS emission.[26] The first-ever published ozone depletion potential (ODP) value assigned to N_2O, 0.017, is less than the ODP value of 1.0 for the reference gas chlorofluorocarbon 11 (CFC-11). While some may not see a cause for alarm based on the ODP value alone, the quantity of N_2O emissions and its potency as a GHG can lead to serious harm (see Table 1).

The ODP value for N2O does not allow for its mandatory inclusion as a class I substance for regulation under the Clean Air Act.[27] However, N_2O could be listed as a class II substance at the direction of the EPA Administrator or regulated under Section 615 of the act.[28] Class I substances have an ODP of 0.2 or more and are more harmful to stratospheric ozone molecules than Class II substances, which have an ODP of less than 0.2.

With or without ODP substance listing, Congress may find it useful to incorporate the ozone depletion impacts of N_2O into its climate change policy proposals both to reduce greenhouse gas emissions and to further ozone recovery achievements. Classifying N_2O emission reduction as an eligible offset type, including N_2O as a covered entity within a cap-and-trade program, or directing EPA to use existing authority under the Clean Air Act to regulate N_2O are other available options to reduce N_2O emissions for ozone or climate protection. Any option chosen to reduce N_2O emissions will more than likely require an improvement of N_2O estimation, measurement, and reporting methods and possible financial incentives.

Congress could apply lessons learned from previous international agreements that are intended to abolish harmful compounds. The outcomes of the Montreal Protocol, put into action in the late 1980s, may prove useful to Congress in understanding the long-term implications of certain climate change policy options, specifically cap-and-trade. A number of gases were phased out under the Protocol, which allowed for each country to establish a regulatory framework to monitor and reduce ODSs. Certain ozone-depleting substances, such as N_2O, were not included in the Protocol partly because their threat was not perceived as urgent at the time. However, one unintended consequence of the success of the Protocol reducing targeted ODSs is that N_2O has emerged as the leading ODS.

End Notes

[1] For more information on CO_2 emission reduction, see CRS Report RL33801, *Carbon Capture and Sequestration (CCS)*, by Peter Folger, and CRS Report RL34621, *Capturing CO_2 from Coal-Fired Power Plants: Challenges for a Comprehensive Strategy*, by Larry Parker and Peter Folger. For more information on legislative proposals to address climate change and regulation of greenhouse gases under the Clean Air Act, see CRS Report R40556, *Market-Based Greenhouse Gas Control: Selected Proposals in the 111th Congress*, by Larry Parker, Brent D. Yacobucci, and Jonathan L. Ramseur; and CRS Report R40585, *Climate Change: Potential Regulation of Stationary Greenhouse Gas Sources Under the Clean Air Act*, by Larry Parker and James E. McCarthy.

[2] U.S. Environmental Protection Agency, *2010 U.S. Greenhouse Gas Inventory Report*, EPA 430-R-10-006, April 2010, http://www.epa.gov/climatechange/emissions/ usinventory report.html.

[3] The potency of a greenhouse gas is described by its global warming potential (GWP), an estimate of how much a greenhouse gas affects climate change over a quantity of time relative to CO_2, which has a GWP value of 1. Intergovernmental Panel on Climate Change, *Climate Change 2007: The Physical Science Basis* (2007), p. 212. The Intergovernmental Panel on Climate Change (IPCC) assigns nitrous oxide and methane a global warming potential of 310 and 21, respectively, over a 100-year time horizon.

[4] An ozone-depleting substance is a compound that contributes to stratospheric ozone depletion by releasing chlorine or bromine atoms into the atmosphere when broken down, leading to the destruction of ozone, a substance necessary to prevent harmful UVB rays from reaching Earth.

[5] The Montreal Protocol is an international treaty constructed to protect the stratospheric ozone layer by gradually eliminating a number of ozone-depleting substances.

[6] Also in the agriculture sector, animal digestive systems and manure management account for a large portion of U.S. methane emissions.

[7] Strategies and technologies for N_2O emission reduction are limited in comparison to resources expended for methane capture. Methane capture technologies, as well as financial and technical support, for point sources have been available for decades. For more information

on methane capture, see CRS Report R40813, *Methane Capture: Options for Greenhouse Gas Emission Reduction*, by Kelsi Bracmort et al.

[8] S. Solomon, D. Qin, and M. Manning et al., *Climate Change 2007: The Physical Science Basis. Contribution of Working Group I to the Fourth*, IPCC, IPCC WG1 AR4 Report, New York, NY, 2007, http://ipcc-wg1.ucar.edu/wg1/wg1-report.html.

[9] Agricultural soil management includes practices that add to, or create an environment conducive to the release of, mineral nitrogen (N).

[10] U.S. Environmental Protection Agency, *2010 U.S. Greenhouse Gas Inventory Report*, EPA 430-R-10-006, April 2010, http://www.epa.gov/climatechange/emissions/usinventoryreport.html.

[11] Mobile (fuel) combustion leads to N2O being emitted as a byproduct. N2O is released as a byproduct during the oxidation of ammonia for production of nitric acid, a primary component of synthetic fertilizers and some explosives. N2O emissions are generally released in large amounts from dry manure handling systems (e.g., pasture, solid storage).

[12] U.S. Department of Agriculture, *U.S. Agriculture and Forestry Greenhouse Gas Inventory: 1990-2005* , Technical Bulletin No. 1921, 2008, http://www.usda. gov/oce/global_change/AFGGInventory1990_ 2005.htm. Nitrogen fixation is the conversion of nitrogen gas to ammonia.

[13] U.S. Environmental Protection Agency, *Global Mitigation of Non-CO2 Greenhouse Gases*, EPA-430-06-005, 2006. Catalytic reduction methods use a catalyst to reduce nitrous oxides in exhaust gas at varying temperatures.

[14] C. S. Snyder, *Fertilizer Nitrogen BMPs to Limit Losses That Contribute to Global Warming* , International Plant Nutrition Institute, Ref. # 08057, June 2008, http://www.ipni.net/ipniweb/portal.nsf/0/6D54ABC2C92D9AFA8525749B0074FF59.

[15] According to the Government Accountability Office, natural gas is the highest-priced factor when producing nitrogen fertilizer. Thus, natural gas prices impact nitrogen fertilizer costs. U.S. General Accounting Office, *Natural Gas: Domestic Nitrogen Fertilizer Production Depends on Natural Gas Availability and Prices*, GAO-03-1148, September 2003, http://www.gao.gov/new.items/d031148.pdf.

[16] For example, the Renewable Fuel Standard (RFS) is a provision established by the Energy Policy Act of 2005 requiring gasoline to contain a minimum amount of fuel produced from renewable biomass (including corn). For more information on the RFS, see CRS Report R40155, *Renewable Fuel Standard (RFS): Overview and Issues*, by Randy Schnepf and Brent D. Yacobucci.

[17] For more information on monitoring carbon in agriculture, see CRS Report RS22964, *Measuring and Monitoring Carbon in the Agricultural and Forestry Sectors*, by Ross W. Gorte and Renée Johnson.

[18] For more information on the agricultural implications of the Mandatory Reporting of Greenhouse Gases Rule, see CRS Report RL32948, *Air Quality Issues and Animal Agriculture: A Primer*, by Claudia Copeland.

[19] For more information on the proposed PSD rule, see CRS Report R40585, *Climate Change: Potential Regulation of Stationary Greenhouse Gas Sources Under the Clean Air Act*, by Larry Parker and James E. McCarthy; and EPA, *Proposed Rule: Prevention of Significant Deterioration and Title V Greenhouse Gas Tailoring Rule*, 2009, http://www.epa.gov/NSR/fs20090930action.html.

[20] For more information on agricultural conservation programs, see CRS Report R40763, *Agricultural Conservation: A Guide to Programs*, by Megan Stubbs. For more information

on technical assistance for nutrient management, see CRS Report RL34069, *Technical Assistance for Agriculture Conservation*, by Megan Stubbs.

[21] For more information on the efforts underway at ARS, visit the Air Quality of Agricultural Systems Research Unit website at http://www.ars.usda.gov/main/site_ main.htm?modecode=36-25-15-15, or the Air Quality National Program website at http://www.ars.usda.gov/research/programs/programs.htm?NP_CODE=203.

[22] U.S. Environmental Protection Agency, *2010 U.S. Greenhouse Gas Inventory Report*, EPA 430-R-10-006, April 2010, http://www.epa.gov/climatechange/emissions/ usinventoryreport.html.

[23].Environmental Protection Agency, Department of Transportation, "Light-Duty Vehicle Greenhouse Gas Emission Standards and Corporate Average Fuel Economy Standards; Final Rule," 75 *Federal Register*, May 7, 2010.

[24] EPA uses the DAYCENT ecosystem model for the U.S. greenhouse gas inventory "to estimate direct N2O emissions from mineral cropland soils that are managed for production of major crops—specifically corn, soybeans, wheat, alfalfa hay, other hay, sorghum, and cotton."

[25] Shilpa Rao and Keywan Riahi, "The Role of Non-CO2 Greenhouse Gases in Climate Change Mitigation: Long-Term Scenarios for the 21st Century," *Energy Journal*, vol. 27 (2006), pp. 1-26; Mario Molina, Durwood Zaelke, and K. Madhava Sarma et al., "Reducing Abrupt Climate Change Risk Using the Montreal Protocol and Other Regulatory Actions to Complement Cuts in CO2 Emissions," *Proceedings of the National Academy of Sciences of the United States of America*, October 12, 2009.

[26] A. R. Ravishankara, John S. Daniel, and Robert W. Portmann, "Nitrous Oxide (N2O): The Dominant Ozone-Depleting Substance Emitted in the 21st Century," *Science Express*, August 27, 2009.

[27] 42 U.S.C. § 7671a. The EPA Administrator may add to the list of class I substances any substances that the Administrator determines as having an ozone depletion potential of 0.2 or greater.

[28] 42 U.S.C. § 7671n. The EPA Administrator has the authority to promptly promulgate regulations respecting the control of an ODS by submitting notice of the proposal and promulgation of such regulation to the Congress.

In: Environmental Considerations in Agriculture ISBN: 978-1-61470-077-7
Editors: E.A. Sargsyan, A.J. Ochion 2011 Nova Science Publishers, Inc.

Chapter 5

THE ROLE OF AGRICULTURE IN REDUCING GREENHOUSE GAS EMISSIONS*

John Horowitz and Jessica Gottlieb

Agriculture could play a prominent role in U.S. efforts to address climate change if farms and ranches undertake activities that reduce greenhouse gas (GHG) emissions or take greenhouse gases out of the atmosphere. These activities may include shifting to conservation tillage, reducing the amount of nitrogen fertilizer applied to crops, changing livestock and manure management practices, and planting trees or grass. The Federal Government is considering offering carbon offsets and incentive payments to encourage rural landowners to pursue these climate-friendly activities as part of a broader effort to combat climate change. The extent to which farmers adopt such activities would depend on their costs, potential revenues, and other economic incentives created by climate policy. Existing Federal conservation programs provide preliminary estimates of the costs of agricultural carbon sequestration.

Figure 1 shows the contributions of broad categories of agricultural and forestry practices and land uses to GHG emissions for the United States in 2007, with red bars representing emissions and green bars representing sequestration, which takes carbon dioxide out of the atmosphere. Changes in these practices and rural land uses can shift both the red and green bars to the left, thereby helping to address climate change. These farming practice and

---

* This is an edited, reformatted and augmented version of the United States Department of Agriculture Economic Research Service publication, dated September, 2010.

land use changes also potentially affect farm profits, meaning that if climate policies aim to involve farmers voluntarily, they would likely need to include payment mechanisms to induce farmers to adopt relevant practices.

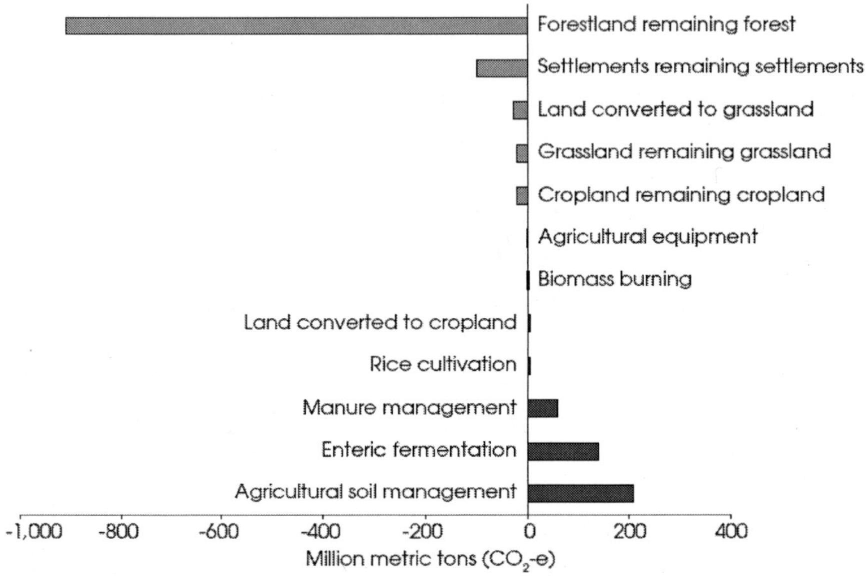

CO_2-e= Equivalent carbon dioxide. CO_2-e is a measure of the global warming caused by all types of greenhouse gases, using the equivalent amount of carbon dioxide (CO_2) as the reference.

Note: A settlement, for our purposes, includes all developed land, including transportation infrastructure and human settlements of any size. Enteric fermentation refers to the microbial fermentation process that occurs in an animal's digestive system and produces methane (CH_4) as a byproduct.

Source: Environmental Protection Agency, Inventory of U.S. Greenhouse Gas Emissions and Sinks: 1990-2007 (April 2009), tables 6-1 and 7-1; agricultural equipment emission source is from tables 3-13 and 3-14.

Figure 1. Agricultural sector GHG emissions and sinks, 2007.

CLIMATE CHANGE POLICIES CAN PROVIDE ROLES FOR AGRICULTURE

The Federal Government is considering a range of actions to reduce GHG emissions. In addition to possible Federal action, State governments and private organizations may also include agriculture as part of their efforts to reduce GHG emissions. The costs, benefits, and overall emission reductions of agricultural mitigation efforts depend on how climate change policy is structured and its stringency.

Carbon Offsets in a Cap-and-Trade System

U.S. efforts to reduce GHG emissions could take the form of a nationwide cap-and-trade system on GHG emissions from the burning of fossil fuels, which account for 80 percent of U.S. emissions. Under this system, fossil fuel sources, such as oil refiners and electricity generators, must have a permit for every ton of emissions. The number of permits that the Government makes available would be the "cap." Under proposed cap-and-trade systems, GHG permits would not be required for emissions arising from agricultural management practices or land use change.

A cap-and-trade system could encourage agricultural mitigation by making agricultural activities eligible for offset credits. Depending on the eligibility rules for offset credits, farmers who increase their use of no-till practices or use nitrogen fertilizer inhibitors, for example, could receive credits for their reduced emissions. They could then sell these credits directly to the industrial sources covered by the cap-and-trade permit requirement or to brokers. Offset credits could be submitted by industrial sources in place of their required emission permits.

Options for Reducing GHG Emissions from Agricultural Enterprises and Increasing Carbon Sequestration on Agricultural Land

Reducing Greenhouse Gas Emissions from Crop and Livestock Production

Livestock and dairy operations emit greenhouse gases through the digestive processes of ruminants (primarily cattle and sheep) and through the decomposition of manure. Livestock managers can reduce methane emissions by changing livestock feeds. Dairy and hog producers can install digesters to capture methane produced during manure storage; the captured methane can then be used to generate electricity. Digesters reduce emissions by converting methane emissions that would have arisen under other methods of manure disposal into less-powerful CO_2 and by generating energy that replaces CO_2 emissions that would have come from fossil fuel based electricity.

Crop producers can change nutrient management practices to reduce emissions from nitrogen fertilizers and manure applied to farmers' fields. Practices that help farmers reduce nitrogen applications without reducing yields include plant tissue testing, soil testing, precision application, use of slow-release fertilizers or nitrification inhibitors, and changes in application timing to better match plant uptake of nutrients (Smith et al., 2008).

Farms and other agricultural enterprises emit CO_2 when they burn gasoline or diesel in vehicles and machinery. If farmers improve operating efficiency or adopt farming techniques that use less fuel, such as no-till, they will reduce their fossil fuel-based GHG emissions. While reducing on-farm fuel use can improve farm profitability and reduce carbon emissions, the potential impact on overall GHG emissions is relatively low because fossil fuel use by agriculture is a small portion of total U.S. fuel consumption.

Producing Renewable Energy on Agricultural Land

Farmers can also contribute to reducing GHG emissions by growing the feedstocks used for biofuels or by installing wind turbines or solar panels on their land. However, indirect land use effects arising from increased demand (in the case of biofuel feedstocks) or reduced supply (in the case of wind or solar power) of crops may lessen or even eliminate the emissions reduction (Fargione et al., 2008).

Increasing Carbon Sequestration on Agricultural Land

To increase carbon sequestration, farmers can adopt practices that increase crop residues left on the field and the volume of organic matter stored in the soil. These practices include switching from conventional tillage to conservation tillage or no-till, reducing or eliminating fallow as part of planned crop rotations, switching from annual to perennial crops, increasing field residues through irrigation, fertilization, planting hay or cover crops, or using additional organic material, such as manure. In some circumstances, these activities present a tradeoff between two greenhouse gases—conservation tillage may increase both carbon sequestration, which mitigates climate change, and nitrous oxide emissions, which contributes to it (Johnson et al., 2007). If these practices reduce yields, then indirect land use changes may also occur.

The box "Options for Reducing GHG Emissions From Agricultural Enterprises and Increasing Carbon Sequestration on Agricultural Land" provides descriptions of agricultural activities that reduce atmospheric concentrations of GHG and could possibly be eligible for offsets. Reductions in fossil fuel use and the production of biofuel feedstocks or other renewable energy would likely not be eligible for offset credits.

Agricultural participation in proposed cap-and-trade markets would be voluntary and, therefore, a potential source of income for farmers. Farmers would be able to decide whether they wanted to undertake activities eligible for offset credits based on the costs of the offset activity and the prices of the offset credits they would earn.

Incentive Payments for Reducing GHG Emissions

Another policy option is to offer direct incentives for agriculture's mitigation activities, an approach similar to current Federal conservation programs, such as the Conservation Reserve Program (CRP). The Government would set rules for eligible practices and payment amounts, just as it does for other conservation programs. To offer incentive payments, the Government would set aside a pool of money or reserve permits to pay farmers for mitigation activities.

Policies to Encourage Renewable Energy Production from Agricultural Sources

A variety of policies, such as the Renewable Fuel Standard, which ensures that transportation fuel sold in the United States contains a minimum volume of renewable fuel, or proposed Renewable Portfolio Standards, which would apply similarly to electricity production, encourage agricultural producers to grow the feedstocks used for biofuels or to provide land for wind or solar power generation.[1]

EXISTING AGRICULTURAL PROGRAMS PROVIDE INSIGHTS INTO POSSIBLE GHG MITIGATION POLICIES AND COSTS

Existing Federal conservation programs provide insights into the costs of agricultural mitigation and yield lessons for the design of greenhouse offset contracts or incentive payments. The CRP and other programs, such as the Environmental Quality Incentives Program (EQIP) and the Wetlands Reserve Program, offer payments for activities similar or identical to the activities that would likely be eligible for GHG offsets.

The Conservation Reserve Program (CRP)

CRP offers payments to farmers to voluntarily retire highly erodible or environmentally sensitive cropland from production for 10-15 years. Enrolled land is then planted to grass or trees, thereby reducing water pollution, improving wildlife habitat, sequestering carbon, and providing other environmental benefits. More than 34 million acres were enrolled in the CRP as of March 2008, either through general or continuous sign-up, making it the largest environmental program on private lands in the United States.[2] The USDA Farm Service Agency (FSA) has estimated that those acres sequestered 48 million more metric tons of CO_2 than if the land had remained in previous uses (FSA, 2008).

Figure 2 shows the average per-acre annual payments that landowners received for general sign-up contracts under the CRP in 2008. Payments are highest overall in the Corn Belt, reflecting the higher opportunity cost of taking land out of crop production there.

Total annualized CRP payments[3] can be combined with FSA estimates of sequestration rates (based in part on Eve et al., 2002) to provide preliminary evidence of the payments farmers would require to sequester carbon through land retirement. The total annual payments that CRP participants receive can also be considered a proxy for the costs they face for retiring cropland; presumably farmers would not participate in the CRP if their costs were substantially higher than the payments they receive. These cost estimates are shown in table 1. Estimated average sequestration costs through tree planting range from $7 to $32 per ton of CO_2 (all references are to metric tons).

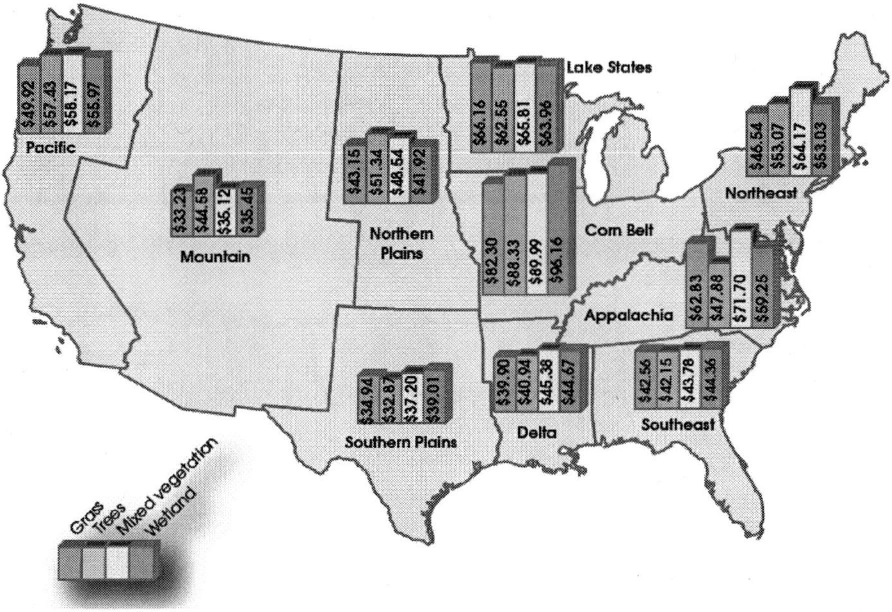

Source: USDA, Economic Research Service calculations based on Conservation Reserve Program data.

Figure 2. Mean annual payment per acre for CRP general sign-up contracts, by vegetation type, March 2008.

Average sequestration costs through grass planting are higher, ranging from $29 to $82 per CO_2 ton, since the forgone agricultural income is roughly the same regardless of vegetation planted (for land enrolled in the CRP) but the amount of CO_2 sequestered is lower.

These costs indicate that if a carbon market were paying $13/ton of CO_2 then owners of the average tree-planted CRP acre in 5 of the 10 regions (Northeast, Appalachia, Southeast, Delta, and Southern Plains) would have found it profitable to enter a carbon market. These estimates are averages, so a range of costs would be exhibited in the individual regions and we would expect at least some tree-planted acres to enter at this price in all regions. By way of comparison, the U.S. Environmental Protection Agency estimates that allowance prices associated with congressional cap-and-trade proposals would range from \$12 to \$41 per ton CO_2 in 2013 and from \$13 to \$59 per ton of CO_2 in 2020.

Actual carbon sequestration levels and costs through land retirement may be lower or higher than these estimates. Only certain kinds of cropland are eligible for CRP payments, so some farms that could provide less costly sequestration may not have been eligible. Furthermore, eligible farms were selected for the program based on many criteria besides their carbon sequestration. These features would likely be changed under any new program that focused on climate benefits, thus providing less expensive carbon sequestration. On the other hand, for climate policies to effectively provide additional carbon sequestration through land retirement, farmers must enroll additional land beyond the current CRP. This effect could cause carbon sequestration costs to be higher than portrayed in table 1.

The Conservation Reserve Program also provides evidence of the acreage that landowners would be likely to offer in a carbon market through individual practices. Figure 3 shows the acres enrolled in general sign-up CRP contracts by vegetation type and region. Grass predominates in all regions except the Southeast and Delta. The small proportion of trees in most regions suggests that farmers were willing to plant trees on only a relatively small proportion of accepted acres at prevailing CRP payment rates.

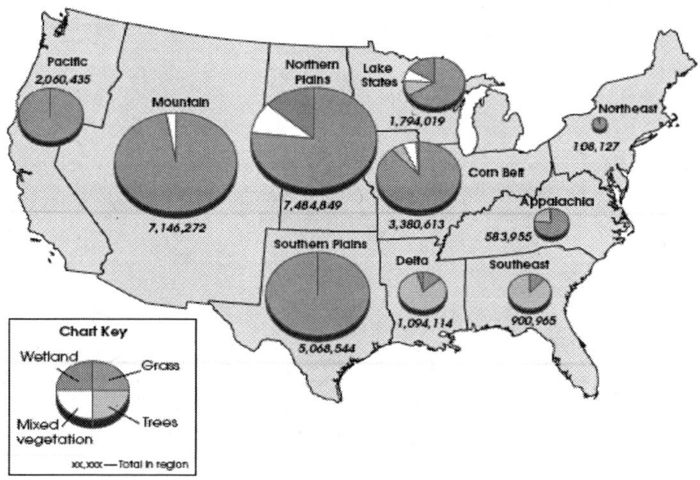

Source: USDA, Economic Research Service calculations based on Conservation Reserve Program data.

Figure 3. Acres enrolled in general sign-up CRP contracts, by vegetation type and region, March 2008.

Table 1. Estimated average costs of carbon sequestration through land retirement, based on CRP enrollment

Region	Average cost per metric ton of CO_2-e (grass)	Average cost per metric ton of CO_2-e (trees)
Northeast	$57	$12
Appalachia	$73	$11
Southeast	$82	$8
Delta	$29	$8
Corn Belt	$68	$32
Lake States	$68	$24
Northern Plains	$43	$16
Southern Plains	$38	$7
Mountain	$55	$20
Pacific Coast	$69	$15

CO_2-e= Equivalent carbon dioxide. CO_2-e is a measure of the global warming caused by all types of greenhouse gases, using the equivalent amount of carbon dioxide (CO_2) as the reference.

Source: USDA, Economic Research Service calculations based on Conservation Reserve Program data.

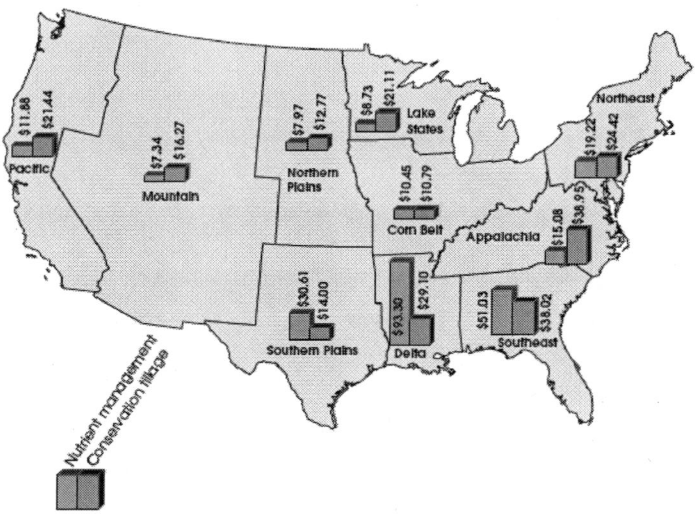

Source: USDA, Economic Research Service calculations based on Conservation Reserve Program data.

Figure 4. Average payment per acre under EQIP contracts, by practice and region, 2008.

This result suggests that substantial numbers of tree-planted acres are likely to be offered to an offset market only if landowners receive higher payments than shown in table 1.

Policy decisions about whether a particular acre would be eligible for CRP enrollment and receive GHG offset credits would affect the owner's willingness to participate in the offset market and/or the CRP, as well as the size of the payment required from each.

Environmental Quality Incentives Program (EQIP)

EQIP provides payments to farmers who adopt a wide range of conservation practices for crop and livestock production. To participate, a producer enters into a contract to implement a specific conservation practice for 1 to 10 years. Per-acre EQIP payments for two of the supported practices that have the potential to contribute to GHG reduction efforts—conservation tillage and nutrient management—are shown in figure 4.

EQIP paid out $42.5 million in 2008 for conservation tillage contracts on 2.7 million acres. If those farms sequestered 0.59 tons of CO_2 per acre, an average estimate, then 1.6 million tons of CO_2 would have been sequestered through conservation tillage, translating to an equivalent outlay of $27 per ton of CO_2. As with the CRP calculations, actual costs of adopting no-till practices under a climate policy may be lower or higher, depending on the success of new policies in bringing in farms that have the lowest carbon sequestration costs but have not yet adopted no-till.

EQIP paid out $35.7 million in 2008 for nutrient management contracts on 4.0 million acres. Many different nutrient management practices are supported by EQIP, so it is not possible to provide a simple estimate of the GHG mitigation resulting from these contracts.

CONCLUSION

U.S. agriculture can contribute to efforts to reduce GHG emissions through increased carbon sequestration, reduced methane and nitrous oxide emissions, and increased renewable energy production. Although conservation programs have been developed and implemented to achieve a host of objectives beyond carbon sequestration, they share several key characteristics of proposed offset markets or other incentives for reducing agricultural GHG emissions, providing a glimpse of the potential farm response to policy incentives for reducing GHG emissions. Further research will develop more precise cost estimates from these programs and for a wider variety of mitigation activities. Forthcoming research will also help identify policy provisions that can encourage cost-effective agricultural efforts and will consider the economic implications of the simultaneous offering of both conservation programs and GHG offsets.

REFERENCES

Eve, M., M. Sperow, K. Howerton, K. Paustian, and R. Follett. "Predicted impact of management changes on soil carbon storage for each cropland region of the coterminous United States," *Journal of Soil and Water Conservation*. 57(4): pp. 196-204, 2002.

Fargione, J., J. Hill, D. Tilman, S. Polasky, and P. Hawthorne. "Land Clearing and the Biofuel Carbon Debt," *Science* 29 319(5867): pp. 1235-38, February 2008.

Johnson, J., A. Franzluebbers, S. Lachnicht Wyers, and D. Reicosky. "Agricultural opportunities to mitigate greenhouse gas emissions," *Environmental Pollution* 150(1): pp. 107-24, 2007.

Smith, P., D. Martino, Z. Cai, D. Gwary, H. Janzen, P. Kumar, B. McCarl, S. Ogle, F. O'Mara, C. Rice, B. Scholes, O. Sirotenko, M. Howden, T. McAllister, G. Pan, V. Romanenkov, U. Schneider, S. Towprayoon, M. Wattenbach, and J. Smith. "Greenhouse gas mitigation in agriculture," *Philosophical Transactions of the Royal Society B* 363: pp. 789-813, February 2008.

U.S. Department of Agriculture, Farm Service Agency. *Conservation Reserve Program: Summary and Enrollment Statistics*, 2008.

U.S. Environmental Protection Agency. *Inventory of U.S. Greenhouse Gas Emissions and Sinks: 1990- 2007*, 2009.

End Notes

[1] See http://www.ers.usda. gov/Briefing/Bioenergy for a discussion of the economic and environmental impacts of biofuel policies.

[2] The Food, Conservation, and Energy Act of 2008 reduced CRP's enrollment cap to 32 million acres as of October 2009.

[3] In addition to the annual rental payments FSA makes to farmers enrolled in the CRP, participants may also be reimbursed for a portion of the costs of establishing a cover crop (e.g., planting grasses or trees). Total average annualized CRP payments are calculated by amortizing the cost share payments over 20 years at 5 percent and adding them to the average annual payments shown in figure 2.

In: Environmental Considerations in Agriculture ISBN: 978-1-61470-077-7
Editors: E.A. Sargsyan, A.J. Ochion 2011 Nova Science Publishers, Inc.

Chapter 6

AGRICULTURAL LAND TENURE AND CARBON OFFSETS*

Roger Claassen and Mitch Morehart

If agricultural producers are allowed to participate in a national cap-and-trade system to curb greenhouse gas (GHG) emissions, the opportunity to sell carbon off sets could prompt farmers to manage their land in a way that increases the amount of carbon stored in soil organic matter and plant biomass, including residue. (See box, "Agriculture and Carbon Off sets.") Farmers who own the land they farm, however, may be in a better position to generate off sets than those who rent their land.

Because carbon is sequestered over time, off set agreements would likely require that sequestration practices be maintained for 5-10 years. For agricultural producers who rent land, insecure land tenure could be a barrier to long-term investment and/or farming practice commitments. Renters may be less interested than landowners in the long-term productivity of the soil and, therefore, less willing than owners to adopt practices that sequester carbon and preserve long-term productivity (e.g., long-term no-till or grass cover), especially if such practices might reduce current returns to the farm.

Th e reluctance of renters to make long-term conservation investments has been observed in a number of previous studies (e.g., Soule et al., 2000). Even if renters are willing to take action, they would be unable to sign long-term

* This is an edited, reformatted and augmented version of the United States Department of Agriculture Economic Research Service publication, dated September, 2009.

carbon off set agreements—or adopt the necessary practices—without a long-term lease.

Landowners who want the fl exibility to change renters or alter rental rates and other contract terms may be reluctant to enter into long-term leases. For example, landowners who rent land for cash may want to consider bids from other producers when crop prices or Government payments rise. Landowners who receive a share of the crop as rent may prefer the option to replace farmers who fail to achieve expected yields. Even if some farmers and landowners work together to generate carbon off sets on rented land, the complexity of the associated lease agreements and the negotiation challenges will continue to act as a deterrent. By contrast, farm operators who own their land will have greater incentives to engage in long-term conservation and will face fewer barriers to signing long-term contracts to generate carbon off sets.

AGRICULTURE AND CARBON OFFSETS

The agriculture and forestry sectors have the ability to remove carbon dioxide from the atmosphere and store ("sequester") it in soils and plant matter. While agriculture is generally excluded from environmental regulations, its ability to offset greenhouse gas (GHG) emissions from regulated sources means agriculture could play an important role in climate policies.

Proposals to combat climate change generally include a national GHG cap-and-trade system that caps the GHG emissions of producers or distributors of fossil fuels and large energy users, but allows these firms to buy and sell excess emissions. Regulated firms could either reduce their pollution levels below the cap or purchase additional emissions allocations from other firms. The ability to trade emissions allows regulated firms to minimize the total cost of pollution control. Costs could be reduced further if regulated firms could buy offsets from unregulated landowners willing to provide low-cost mitigation through soil and biomass sequestration.

While agricultural participation in a cap-and-trade market might be cost effective, implementation poses major challenges. The extent to which farmers would be willing to participate depends on a host of program design issues, including eligibility criteria, baseline standards (below which credits would not be generated), allowable interactions with USDA conservation programs, and farm characteristics and offset prices. This publication examines farms that may be in the best position to participate in a cap-andtrade system.

Could land tenure be a limiting factor in generating agricultural carbon offsets? The answer depends, in part, on the intersection of land tenure patterns and carbon sequestration potential. If a large proportion of farmers who own all or most of the land they farm are willing to offer extensive opportunities to generate carbon emission offsets, tenure is less likely to be a limiting factor. This report assesses the intersection of land tenure and carbon sequestration potential based on an assumed scenario for generating carbon offsets. (See box, "Carbon Sequestration Potential: Definition, Data, and Methods.") While we do not know how large carbon offset payments will be or how individual farm operators will respond to them, a rough approximation of the distribution of potential carbon sequestration is possible based on current land uses and practices. Whether land tenure will actually be a limiting factor can be assessed only in the context of future market demand for offsets and the ability of agriculture to generate offsets within the legal framework that may be established for carbon offset trading.

LAND TENURE AND CARBON SEQUESTRATION POTENTIAL

We separate farms and ranches into four tenure classes (Figure 1). High-tenure operators, for example, own at least 80 percent of the land they farm or ranch; low-tenure operators own less than 20 percent.

Carbon sequestration in U.S. agriculture (for offsets or other purposes) will be determined by (1) the amount of carbon that can be sequestered per acre through adoption of carbon sequestering practices and (2) the number of acres on which farmers are willing to adopt such practices. Carbon sequestration potential diff ers across practices and regions and depends on previous land use. On average, conversion of cropland to permanent cover (e.g., grass or trees) yields larger per acre increases in sequestered carbon than do changes in production practices (e.g., conservation tillage) on cultivated cropland. As a source of carbon sequestration off sets, land-use change may also be easier to confi rm and enforce than changes in tillage or other crop production practices.

About two-thirds of estimated carbon sequestration potential comes from landuse change: shifting cultivated cropland to grass cover for conserving use (e.g., Conservation Reserve Program (CRP)) or for use as hay or pasture. On high-tenure farms, three-quarters of sequestration potential is derived from land-use change (figure 1).

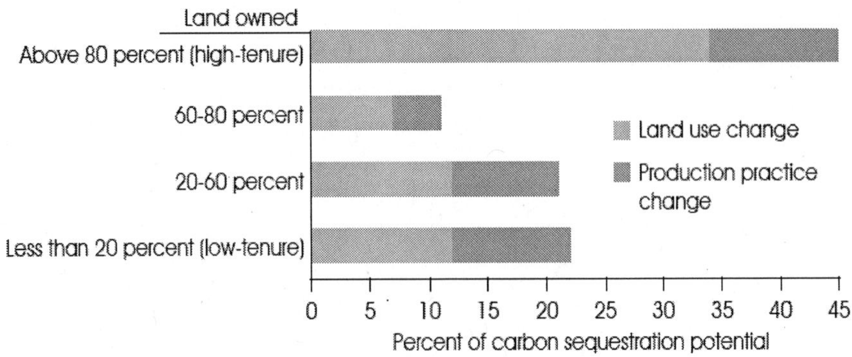

Note: Percentages do not add to 100 because of rounding.

Source: ERS tabulations based on USDA's 2002 Agricultural Resource Management Survey, conducted by the National Agricultural Statistics Service and the Economic Research.Service, and Eve et al., 2002.

Figure 1. Potential carbon sequestration, by land tenure and practice type.

Given the correlation between high-tenure farms and land in permanent grass cover or CRP, high-tenure farms are generally assigned higher carbon sequestration potential. In this simple simulation, high-tenure farms account for about 45 percent of the sector's carbon sequestration potential (Figure 1).

An examination of the intersection between land tenure and farming practice patterns shows why high-tenure farms may be in the best position to adopt carbon-sequestering practices. These farms and ranches maintain a relatively large proportion of land in hay and pasture (fig. 2). Tis land-use pattern may signal willingness to pursue low-intensity production consistent with carbon sequestration goals. Unlike producers in other land tenure categories, these farmers and ranchers focus more on livestock production and account for more than half of all livestock sales (fig. 3). Tese farms may be in the best position to utilize additional pasture and hay land created by carbon sequestration incentives. High-tenure farms are also more likely than other farms to enroll land in the CRP, accounting for about two-thirds of CRP acreage (fig. 2). Because it provides fi nancial incentives for farmers and landowners to establish conservation cover on cultivated land and to maintain the cover for at least 10 years, CRP may be a good barometer of willingness to participate in carbon off set markets. CRP rules also require landowners to off er tenants the opportunity to participate in the CRP contract and receive CRP payments.

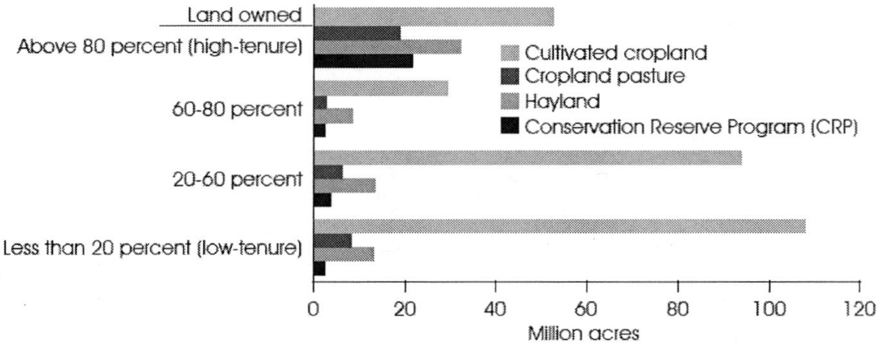

Note: Percentages do not add to 100 because of rounding.

Source: ERS tabulations based on USDA's 2002 Agricultural Resource Management Survey, conducted by the National Agricultural Statistics Service and the Economic Research Service, and Eve et al., 2002.

Figure 2. Land use, by tenure class.

In contrast to the high-tenure group, farms in the two low-tenure groups (those owning 0-20 percent and 20-60 percent of the land they operate) are far more likely to use land in cultivated crop production and less likely to enroll land in the CRP (Figure 2).

Farms in these two tenure classes account for more than 70 percent of cultivated cropland and more than 60 percent of crop production (by value). Low-tenure farms include roughly 40 percent of hay and cropland pasture acreage, roughly 40 percent of livestock production (by value), and only 20 percent of CRP acres. T ese farms tend to focus more on crop production and are reluctant to retire land rather than farm it.

Carbon Sequestration Potential: Definition, Data, and Methods

Carbon sequestration potential is the amount of carbon that could be sequestered (held) within farmland in response to financial incentives generated by a GHG offset market. In the broadest sense, carbon sequestration potential can be estimated as the total number of farmland acres where emission reduction practices could be applied, multiplied by an estimate of the amount of carbon sequestration per acre. For our purposes, we assume that, at least initially, incentives will induce relatively small or *marginal* changes in land use and production practices on individual farms and ranches.

Specifically, we assume that all farmers increase their use of carbon sequestering practices by 10 percent. For example, a producer who has 20 percent of land in pasture would increase grazing acreage to 22 percent by shifting land from crop production. A simple index value for potential carbon sequestration can be created by multiplying the shifted acreage by the coefficients reported in table 1. Similar values can be calculated for marginal (10 percent) changes in other carbon sequestering practices. By summing all of these calculations, we arrive at an overall index of carbon sequestration potential. This index is intended for illustrative purposes only; actual land use or production practice changes will depend on a host of economic, policy, and bio-physical factors.

High levels of sequestration are reported for the conversion of cultivated cropland to grass or tree cover, relative to changes in cropland management in table 1. Land-use data, based largely on the implementation of farm commodity programs, also may make it easy to confirm that offsets are being generated on land that was previously in crop production, at least for the 75 percent of U.S. cropland on farms that receive Federal farm commodity program payments.

Reported coefficients also show larger carbon sequestration for pasture and hayland than for land enrolled in the Conservation Reserve Program (CRP), under the assumption that CRP lands are never harvested. CRP land can, however, be used for hay or grazing under certain conditions, possibly increasing carbon sequestration. Hay and pasture may also receive fertilizer, potentially creating additional GHG emissions. Nonetheless, our larger point remains valid—land-use change is likely to yield larger and more readily verifiable GHG reduction than changes in crop production practices.

Data on land use and farm practices come from the Agricultural Resources Management Survey (ARMS) conducted jointly by USDA's Economic Research Service and National Agricultural Statistics Service. The ARMS data are based on an annual survey of 34,000 farms that elicits information on farm businesses and farm households. Farm-specific weights can be used to expand estimates to the full farm population. For each observation (surveyed farm), ARMS data include information on land use, land tenure, CRP participation, production practices, and production of crop and livestock commodities. The 2002 ARMS data are used because data from that year provide a full accounting of agricultural land use, including pasture and rangeland.

Table 1. Descriptive statistics for carbon sequestration coefficients

	Appalachian States	Corn Belt	Delta States	Lake States	Mountain States	North-east	Northern Plains	Pacific States	South-east	Southern Plains
				Metric tons of carbon per acre per year						
Change from conventional tillage to reduced tillage	0.06	0.09	0.11	0.07	0.06	0.06	0.06	0.05	0.04	0.06
Change from conventional tillage to no tillage	0.13	0.17	0.21	0.15	0.13	0.13	0.13	0.10	0.08	0.13
Increased residue through use of manure or cover crops	0.13	0.17	0.21	0.15	0.08	0.13	0.13	0.10	0.08	0.13
Change from cultivated summer fallow to continuous cropping	0.00	0.00	0.00	0.15	0.08	0.00	0.13	0.10	0.00	0.13
Retire land from cultivated crop production (CRP)	0.19	0.25	0.30	0.21	0.12	0.19	0.19	0.14	0.11	0.18
Conversion of cultivated cropland to pasture or hay	0.38	0.50	0.59	0.42	0.23	0.38	0.37	0.29	0.23	0.36

Source: Eve, M.D., M. Sperrow, K. Howerton, K. Paulson, and R.F. Follet. "Predicted Impact of Changes on Soil Carbon Storage for Each Cropland Region of the Conterminous United States," Journal of Soil and Water Conservation 57,4 (July-August 2002): 196-204.

The propensity of high-tenure farms toward hay, pasture, and CRP land use may be explained in a number of ways. First, these farms may be on land that is less productive when used for crops. Low-productivity land could also explain why high-tenure farms focus on livestock production that can effectively utilize pasture and hay, relative to other tenure classes (figure 3). Land productivity could also explain their willingness to enroll formerly cultivated land in the CRP. An alternate explanation is that small, rural residence farms, which account for 72 percent of all high-tenure farms (Figure 4), are engaged in agricultural production only as a side venture and are more likely than other farms to be engaged in cow-calf operations (Hoppe et al., 2007). Such farms may be more willing than others to place land in retirement (CRP) or low-intensity use (pasture or hay) because their income derives mostly from off the farm.

From a carbon sequestration perspective, however, the largest share of sequestration potential is on large farms. Half of the total carbon sequestration potential is concentrated on just 7 percent of farms (high and medium-high carbon potential farms in Figure 5) that tend to be located in the Northern Plains and Mountain regions. While small in number, these farms operate 38 percent of all farmland. On average, they operate three times the acreage of the remaining 93 percent of farms that account for the other half of potential carbon sequestration. This "high sequestration potential" group is similar to (and overlaps with) the high-tenure group in that it accounts for a large share of hay and cropland pasture (52 percent) and CRP enrollment (58 percent). As a group, its relatively high participation in CRP demonstrates a willingness to participate in environmental land-use change programs.

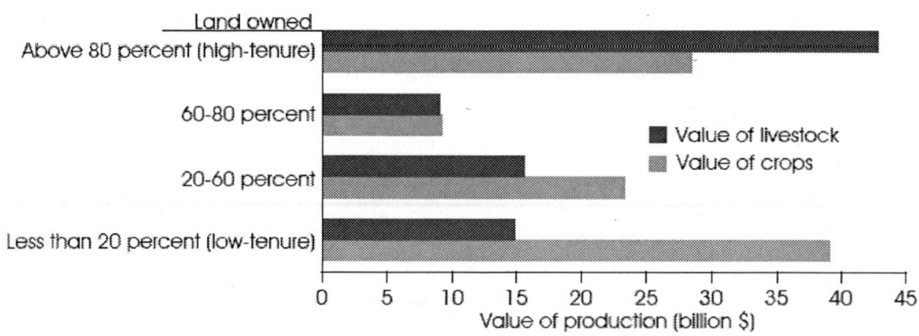

Source: ERS tabulations based on USDA's 2002 Agricultural Resource Management Survey, conducted by the National Agricultural Statistics Service and the Economic Research Service.

Figure 3. Value of livestock and crop production, by tenure class.

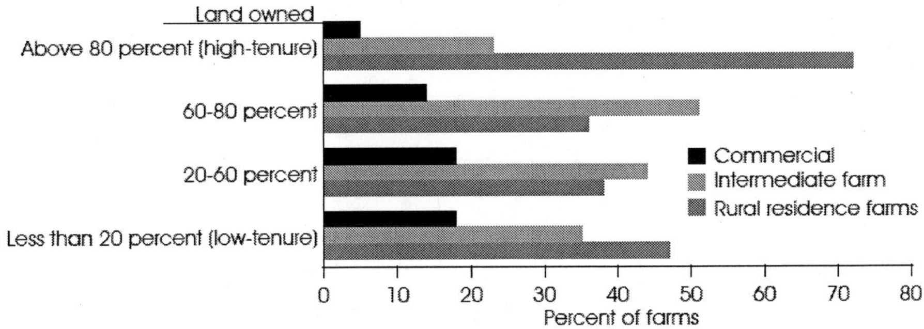

Source: ERS tabulations based on USDA's 2002 Agricultural Resource Management Survey, conducted by the National Agricultural Statistics Service and the Economic Research Service.

Figure 4. Farms by tenure and typology.

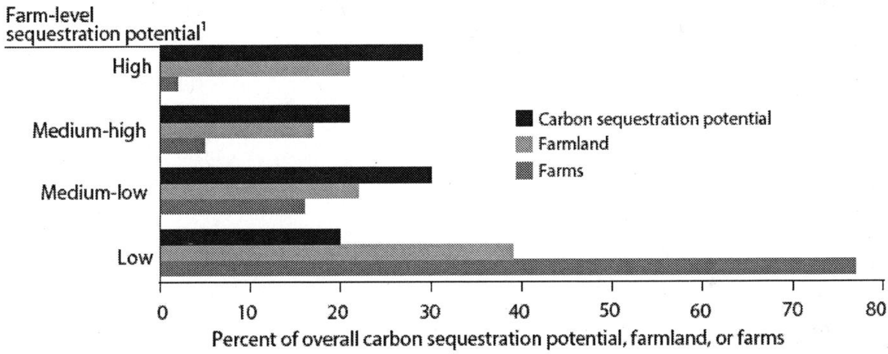

[1] Individual farms, ranked by their carbon sequestration potential. The top set of bars indicates that the 2 percent of farms with the highest level of sequestration potential account for roughly 30 percent of the agricultural sector's total carbon sequestration potential, as measured in this simulation.

Source: ERS tabulations based on USDA's 2002 Agricultural Resource Management Survey, conducted by the National Agricultural Statistics Service and the Economic Research Service.

Photo credit: Jupiterimages.

Figure 5. Carbon sequestration potential, farms, and farmland, by farm-level sequestration potential.

Conclusions

If the United States adopts a national cap-and-trade system to curb GHG, and if agricultural producers are allowed to sell carbon off sets, participation is likely to require long-term commitments. Land ownership could play a role in the participation decisions of agricultural producers. A large share of agricultural carbon sequestration potential is on high-tenure farms. As a group, these farms are more likely to focus on livestock production. They may also be more willing than nonlivestock farms to convert marginal cropland to grazing use. High-tenure farms are also more likely to enroll land in CRP, direct evidence of their willingness to accept payments for land use change. While it is not yet known whether or to what extent farmers will be eligible to participate in a carbon off set market, previous land-use and farming practice decisions suggest that high-tenure farm operators will account for a large part of the agricultural sector's carbon sequestration potential. If so, distribution of farmland ownership should not be a limiting factor in the sector's participation in climate change programs for the foreseeable future.

The analysis presented here is illustrative. We assumed that the price signals from a national cap-and-trade system would encourage farm operators to make marginal changes in their operations, but would not encourage nonoperator landlords to make radical changes in their rental arrangements. If carbon off set prices are high enough (relative to applicable agricultural commodity prices), landowners may choose to alter their leasing arrangements, taking land out of production or placing restrictions on the farming practices operators can adopt. In this case, the sector's carbon sequestration potential could be more broadly distributed than this analysis indicates. Additional research is needed to better quantify the role of land tenure in producer response to carbon sequestration-related incentives for land use and production practice change.

References

Hoppe, Robert A., Penni Korb, Erik J. O'Donoghue, and David E. Banker. *Structure and Finances of U.S. Farms: Family Farm Report, 2007 Edition*, EIB24, p. 9, U.S. Department of Agriculture, Economic Research Service, June 2007.

Soule, M.J., A. Tegene, and K.D. Wiebe. "Land Tenure and the Adoption of Conservation Practices," *American Journal of Agricultural Economics* 82:993-1005, 2000.

CHAPTER SOURCES

Chapter 1 - This is an edited, reformatted and augmented version of a Congressional Research Service publication, R41622, dated February 7, 2011.

Chapter 2 - This is an edited, reformatted and augmented version of Congressional Research Service publication, R41530, dated December 10, 2010.

Chapter 3 - This is an edited, reformatted and augmented version of a Congressional Research Service publication, RL33898, dated November 9, 2009.

Chapter 4 - This is an edited, reformatted and augmented version of a Congressional Research Service publication, R40874, dated January 4, 2011.

Chapter 5 - This is an edited, reformatted and augmented version of a United States Department of Agriculture Economic Research Service publication, dated September, 2010.

Chapter 6 - This is an edited, reformatted and augmented version of a United States Department of Agriculture Economic Research Service publication, dated September, 2009.

INDEX

A

B

D

E

G

H

I

O

P

T